Special thanks to the authors, editors, art directors,

copyeditors, and other staff members of *Fine Homebuilding*

who contributed to the development of the articles in this book.

CONTENTS

 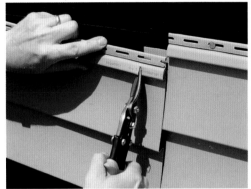

PART 3 : FRAMING ROOFS

PART 4 : SIDING

PART 5 : FINISH CARPENTRY

PART 6 : WINDOWS AND DOORS

PART 7 : STAIRS

PART 8 : CABINETS AND BUILT-INS

INTRODUCTION

Like a lot of baby boomers, I grew up in a house with a workbench in the basement and a pegboard full of carpentry tools hanging behind it. My father had been raised on a farm during the Depression, and even though he became a successful surgeon, he would no more live in a house without tools than he would drive a car without a flashlight in the glove box. I also went to a public school that taught industrial arts. My mother still uses the walnut doorstop shaped like a Scottie dog that I made when I was 12.

My early exposure to the satisfaction of building things led me to spend the first ten years of my professional life working as a carpenter. To this day, it's still how I define myself. Ask me what I do, and I'll say I'm an editor. But ask me who I am, and I'll say I'm a carpenter.

Unfortunately there are fewer homes today with workbenches in the basement and fewer fathers who can teach their kids how to build things. That's where this book can help. Collected here are 37 articles from past issues of *Fine Homebuilding* magazine. Written by professional carpenters from all over the country, these articles represent an apprenticeship in the trade. Whether you're fixing up your own house on the weekends or earning a living building homes for other people, you'll find useful advice and hard-won tips in this book. Heck, you'll even learn how to build a set of shelves to put it on.

—Kevin Ireton, editor,
Fine Homebuilding

10 Rules for Framing

■ BY LARRY HAUN

It was a coincidence that another contractor and I began framing houses next door to one another on the same day. But by the time his house was framed, mine was shingled, wired, and plumbed. It was no coincidence that the other contractor ran out of money and had to turn the unfinished house over to the lending company, while I sold mine for a profit.

Both houses were structurally sound, plumb, level, and square, but every 2x4 in the other house was cut to perfection. Every joint looked like finish carpentry. The other contractor was building furniture, and I was framing a house.

Unlike finish carpentry, framing doesn't have to look perfect or satisfy your desire to fit together two pieces of wood precisely. Whether you're building a house, an addition, or a simple wall, the goals when

framing are strength, efficiency, and accuracy. Following the building codes and the blueprints should take care of the strength; efficiency and accuracy are trickier. But during 50 years of framing houses, I've come up with the following rules to help me do good work quickly and with a minimum of effort.

Larry Haun, author of The Very Efficient Carpenter *(The Taunton Press, 1992) and* Habitat for Humanity How to Build a House *(The Taunton Press, 2002 —A revised and expanded version of* Habitat for Humanity How to Build a House *is due to publish in 2008), has been framing houses for more than 50 years. He lives in Coos Bay, Oregon.*

1. Don't move materials any more than you have to.

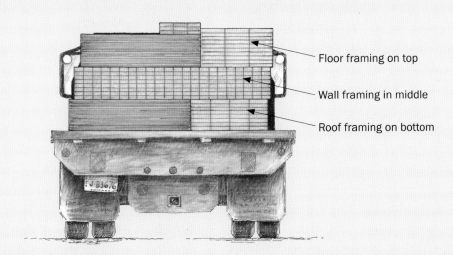

Floor framing on top

Wall framing in middle

Roof framing on bottom

Hauling lumber from place to place is time-consuming and hard on your body. Make it easier on yourself every chance you get, and start by having the folks at the lumberyard do their part. Make sure lumber arrives on the truck stacked in the order it will be used. You don't want to move hundreds of wall studs to get to your plate stock, for instance. And floor joists go on top of floor sheathing, not the other way around.

When it's time for the delivery, unload the building materials as close as possible to where they will be used. Often lumber can be delivered on a boom truck, so stacks of lumber can be placed right up on the deck or on a simple structure built flush alongside the deck.

Once the material is delivered, don't move it any more than you need to. Cut studs, plywood, and anything else you can right on the stack. If you do have to move wood, plan so that you have to move it only once.

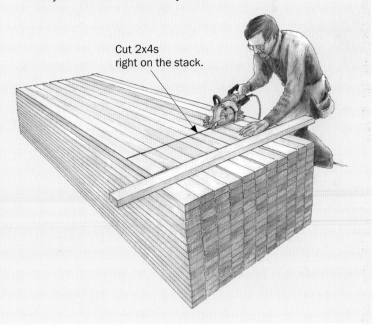

Cut 2x4s right on the stack.

2. Build a house, not furniture.

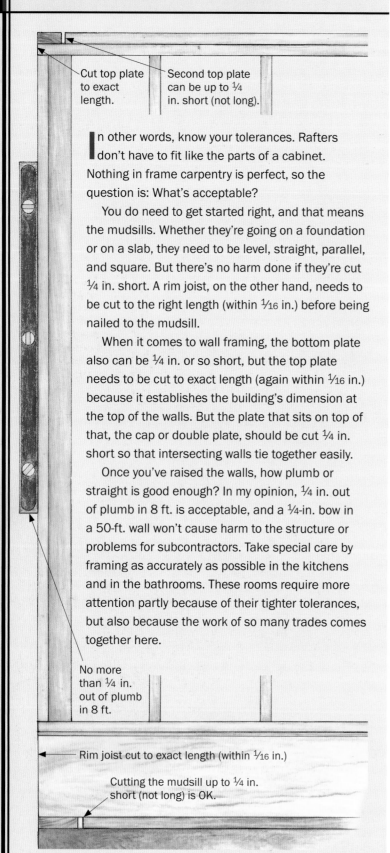

Cut top plate to exact length.

Second top plate can be up to ¼ in. short (not long).

In other words, know your tolerances. Rafters don't have to fit like the parts of a cabinet. Nothing in frame carpentry is perfect, so the question is: What's acceptable?

You do need to get started right, and that means the mudsills. Whether they're going on a foundation or on a slab, they need to be level, straight, parallel, and square. But there's no harm done if they're cut ¼ in. short. A rim joist, on the other hand, needs to be cut to the right length (within ¹⁄₁₆ in.) before being nailed to the mudsill.

When it comes to wall framing, the bottom plate also can be ¼ in. or so short, but the top plate needs to be cut to exact length (again within ¹⁄₁₆ in.) because it establishes the building's dimension at the top of the walls. But the plate that sits on top of that, the cap or double plate, should be cut ¼ in. short so that intersecting walls tie together easily.

Once you've raised the walls, how plumb or straight is good enough? In my opinion, ¼ in. out of plumb in 8 ft. is acceptable, and a ¼-in. bow in a 50-ft. wall won't cause harm to the structure or problems for subcontractors. Take special care by framing as accurately as possible in the kitchens and in the bathrooms. These rooms require more attention partly because of their tighter tolerances, but also because the work of so many trades comes together here.

No more than ¼ in. out of plumb in 8 ft.

Rim joist cut to exact length (within ¹⁄₁₆ in.)

Cutting the mudsill up to ¼ in. short (not long) is OK.

3. Use your best lumber where it counts.

These days, if you cull every bowed or crooked stud, you may need to own a lumber mill to get enough wood to frame a house. How do you make the most of the lumber that you get?

Use the straightest stock where it's absolutely necessary: where it's going to make problems for you later on if it's not straight. Walls, especially in baths and kitchens, need to be straight. It's not easy to install cabinets or tile on a wall that bows in and out. And straight stock is necessary at corners and rough openings for doors.

The two top wall plates need to be straight as well, but the bottom plate doesn't. You can bend it right to the chalkline and nail it home. If you save your straight stock for the top plates, you'll have an easy time aligning the walls. And every project needs lots of short stock for blocking; take your bowed material and cut it into cripples, headers, and blocks.

4. Work in a logical order.

Establish an efficient routine for each phase of work, do it the same way every time, and tackle each phase in its logical order. In the long run, having standard procedures will save time and minimize mistakes. Let's take wall framing as an example.

First I snap all of the layout lines on the floor; then I cut the top and bottom plates and tack all of them in place on the lines. Next I lay out the plates, detailing the location of every window, door, stud, and intersecting wall.

I pry up the top plate and move it about 8 ft. away from the bottom plate, which I leave tacked to the deck. I scatter studs every 16 in. for the length of the wall. I nail the top plate to the studs and keep the bottom of the studs snug against the bottom plate. This helps to keep the wall square, straight, and in position to be raised. I try to establish a rhythm and work consistently from one end to the other. Once the top plate is completely nailed, I pry up the bottom plate and repeat the process on the bottom.

It's worth saying that I didn't just make up these steps; they evolved over time. Recognizing inefficiency is an important part of framing.

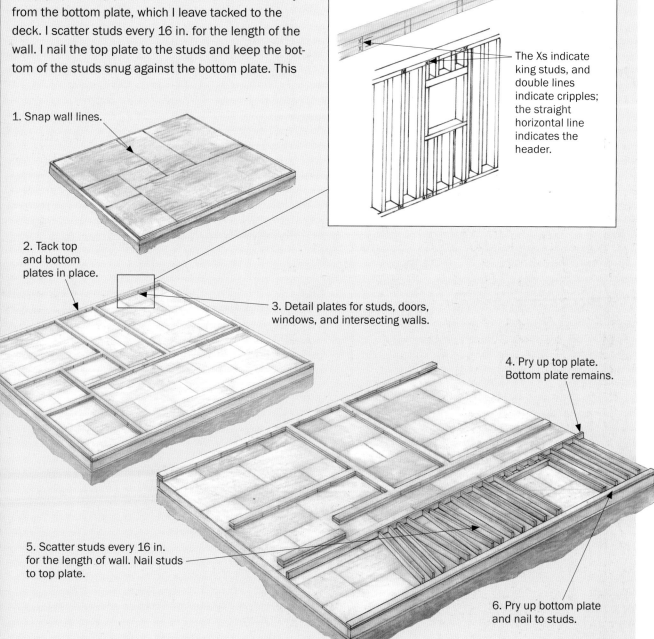

The Xs indicate king studs, and double lines indicate cripples; the straight horizontal line indicates the header.

1. Snap wall lines.

2. Tack top and bottom plates in place.

3. Detail plates for studs, doors, windows, and intersecting walls.

4. Pry up top plate. Bottom plate remains.

5. Scatter studs every 16 in. for the length of wall. Nail studs to top plate.

6. Pry up bottom plate and nail to studs.

5. Keep the other trades in mind.

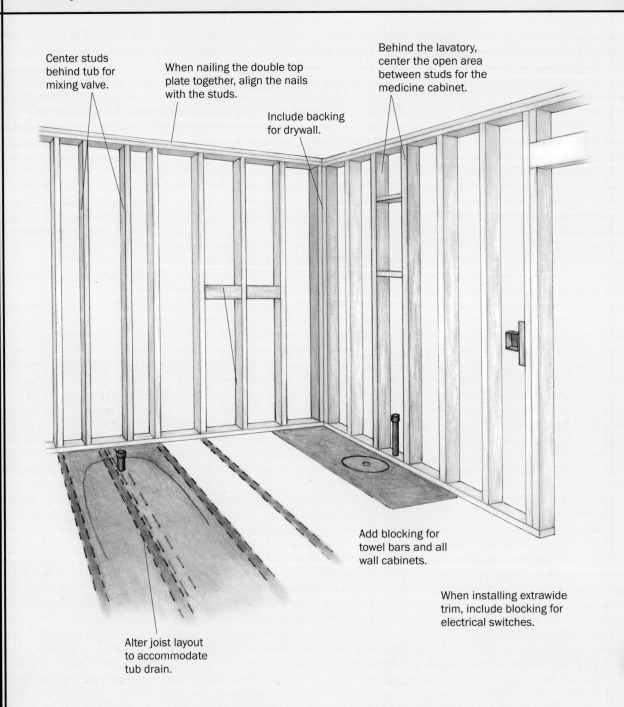

Center studs behind tub for mixing valve.

When nailing the double top plate together, align the nails with the studs.

Include backing for drywall.

Behind the lavatory, center the open area between studs for the medicine cabinet.

Add blocking for towel bars and all wall cabinets.

When installing extrawide trim, include blocking for electrical switches.

Alter joist layout to accommodate tub drain.

If you want to waste time and money when framing, don't think about the electrical work, the plumbing, the heat ducts, the drywall, or the finish carpentry. Whether you do them yourself or hire subcontractors, these trades come next. And unless you're working with them in mind every step of the way, your framing can be in the way.

For example, when you nail on the double top plate, keep the nails located over the studs. This tip leaves the area between the studs free for the electrician or plumber to drill holes without hitting your nails.

6. Don't measure unless you have to.

The best way to save time when you're framing a house is by keeping your tape measure, your pencil, and your square in your nail pouch as much as possible. I have to use a tape measure to lay out the wall lines accurately on the deck, but after that, I cut all of the wall plates to length by cutting to the snapped wall lines. I position the plate on the line, eyeball it, and then make the cuts at the intersecting chalkline.

Another time-saver is to make square crosscuts on 2x4s or 2x6s without using a square. Experience has shown me that with a little practice, anyone can make these square cuts by aligning the leading edge of the saw's base, which is perpendicular to the blade, with the far side of the lumber before making the cut.

Trimming ¼ in. from a board's length shouldn't require measuring. Ripping (lengthwise cuts) longer pieces also can be done by eye if you use the edge of the saw's base as a guide. Train your eye. It'll save time cutting, and as you develop, you'll also be able to straighten walls as easily by eye as with a string.

With practice, you can make square cuts by aligning the front edge of the saw's base with the far edge of the board.

7. Finish one task before going on to the next.

My first framing job was with a crew that would lay out, frame, and raise one wall at a time before moving on to the next. Sometimes they would even straighten and brace the one wall before proceeding. We wasted a lot of time constantly switching gears.

If you're installing joists, roll them all into place and nail them before sheathing the floor. Snap all layout lines on the floor before cutting any wall plates, then cut every wall plate in the house before framing. If you're cutting studs or headers and cripples, make a cutlist for the entire project and cut them all at once. Tie all the intersecting walls together before starting to straighten and brace the walls.

Finishing before moving on is just as important when it comes to nailing and blocking. You might be tempted to skip these small jobs and do them later, but don't. Close out each part of the job as well as you can before moving on to the next. Working in this way helps to maintain momentum, and it prevents tasks from being forgotten or overlooked.

8. Cut multiples whenever possible.

You don't need a mathematician to know that it takes less time to cut two boards at once than it does to cut each one individually.

If you have a stack of studs that all need to be cut to the same length, align one end of the top row, snap a chalkline all the way across, and cut the studs to length right on the pile. Or you can spread them out on the floor, shoving one end against the floor plate, snap a chalkline, and cut them all at once.

Joists can be cut to length in a similar way by spreading them out across the foundation and shoving one end up against the rim joist on the far side. Mark them to length, snap a line, and cut the joists all at once.

Also, don't forget to make repetitive cuts with a radial-arm or chop/miter saw outfitted with a stop block, which is more accurate and faster than measuring and marking one board at a time.

First, spread studs on the plywood floor with one end against the floor plate.

Then mark them and cut along the chalkline.

9. Don't climb a ladder unless you have to.

I don't use a ladder much on a framing job except to get to the second floor before stairs are built. Walls can be sheathed and nailed while they're lying flat on the deck. Waiting until the walls are raised to nail on plywood sheathing means you have to work from a ladder or a scaffold. Both are time-consuming.

With a little foresight, you can do the rafter layout on a double top plate while it's still on the floor. Otherwise, you'll have to move the ladder around the job or climb on the walls to mark the top plate.

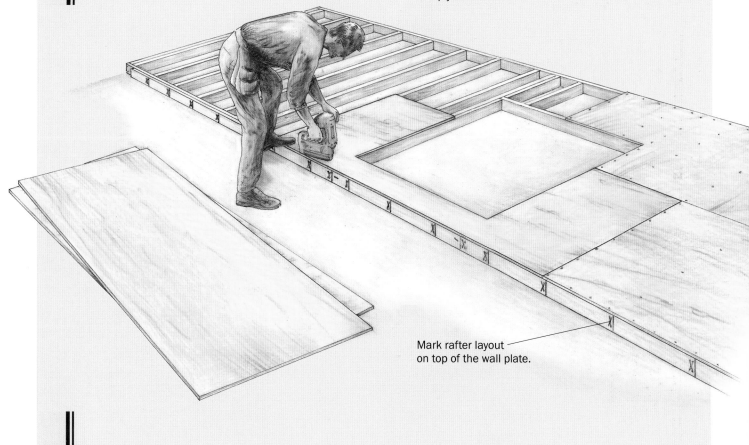

Attach the sheathing while the wall is still on the plywood floor.

Mark rafter layout on top of the wall plate.

10. Know the building code.

Building codes exist to create safe structures. Because building inspectors are not capable of monitoring all parts of every project, it's your responsibility to know the building code and to build to it.

For instance, the code actually specifies how to nail a stud to a wall plate. You need two 16d nails if you're nailing through a plate into the end of the stud, or four 8d nails if you're toenailing. When you nail plywood or oriented strand board (OSB) roof sheathing, you need a nail every 6 in. along the edge of the sheathing and every 12 in. elsewhere. And if you're using a nail gun, be careful not to overdrive the nails in the sheathing.

And a final word: If special situations arise, consult the building inspector. He or she is your ally, not your enemy. Get to know the building code for your area. Get your own copy of the IRC (International Residential Code) and build well, but build efficiently, with the understanding that perfection isn't what is required.

Roof sheathing is nailed every 6 in. along the edges and every 12 in. elsewhere. In high-wind areas, sheathing along the eaves, rakes, and ridges is nailed every 6 in.

$\frac{5}{8}$-in. sheathing

2x8 rafter

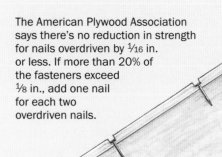

The American Plywood Association says there's no reduction in strength for nails overdriven by $\frac{1}{16}$ in. or less. If more than 20% of the fasteners exceed $\frac{1}{8}$ in., add one nail for each two overdriven nails.

11. Work safely whatever the rule.

Working safely should be at the top of your priority list. Safety glasses, hearing protection, and a dust mask should be the norm, as should attention around coworkers or dangerous debris.

Safety devices and good intentions, however, won't help if your mind isn't on the work. Pay attention, approach the work with a clear head, listen to that inner voice that says, "This is too dangerous," and be extra careful toward the end of the day. —L. H.

Avoiding Common Framing Errors

■ BY RICK TYRELL

When it comes to building inspectors, I'm like the teenager who thought his father was a poor simpleton who knew nothing about anything. A few years later, after the son had a child of his own, he was amazed at how much the old man had learned in such a short time. The truth is, I'm amazed by how much building inspectors have learned since I began working in construction 12 years ago.

As a builder/remodeler, I sometimes thought that building inspectors pushed points in the code that didn't seem all that important. But after years of learning—from my own mistakes and the mistakes of others—I now understand that the code really provides only minimum requirements for health and safety. It's important to remember that if there is a code violation, the builder assumes all responsibility, and the building official assumes none.

Remember, too, that the importance of some code issues isn't obvious until the integrity of a structure is tested by some sort of natural disaster. Hurricanes are perfect examples. Most people in a storm's path aren't aware of any code violations until their buildings collapse during the storm. Improperly secured plywood blows off roofs. Some buildings blow off their foundations because of inadequate anchoring.

The House Must Be Anchored Firmly to Its Foundation

According to the code, sill plates must provide a minimum protection against termite and decay damage and must be of pressure-treated lumber or naturally rot-resistant wood such as heartwood of redwood, black locust, or cedar.

Before I knew better, I used pressure-treated lumber that was treated to 0.25, which refers to the retention level of the wood preservatives that are forced into the wood. However, I should have used lumber that had a retention level of 0.40. According to the Southern Forest Products Association, 0.40-retention, or ground-contact, lumber is required anytime treated lumber comes in contact with concrete that contacts the ground. To know if lumber has been treated

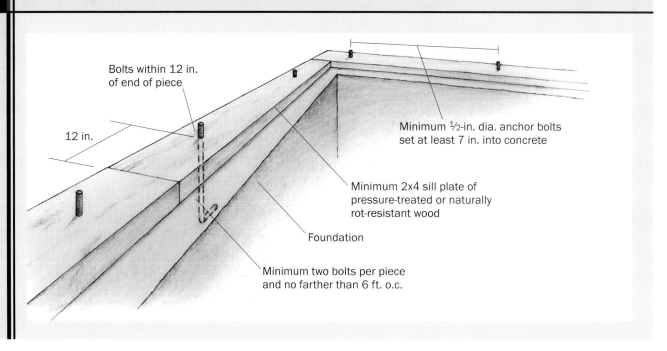

Bolts within 12 in. of end of piece

12 in.

Minimum ½-in. dia. anchor bolts set at least 7 in. into concrete

Minimum 2x4 sill plate of pressure-treated or naturally rot-resistant wood

Foundation

Minimum two bolts per piece and no farther than 6 ft. o.c.

to level 0.40, read the grade stamp on the lumber or the tag stapled to the lumber.

Once the sills are in place, they must be anchored to the foundation. According to the Council of American Building Officials' (CABO) model code, anchor bolts may not be placed less than 7 in. into concrete, must be a minimum of ½ in. in dia. and must be spaced no farther than 6 ft. o.c. Also, each section of sill must have at least two bolts that go through a minimum 2x4 stock, and the bolts must be placed within 12 in. of the ends of the sills.

Top Plates Require Offset Joints and Adequate Fastening

I remember photographs of damage inflicted by hurricane Hugo. One photo showed a wall that had let go because a joint in the double top plate was inadequately offset from another. The whole wall gave way between those two joints.

Improper fastening and offsetting of double-top-plate joints are the most common errors found in bearing-wall construction. By code, the double plate must overlap the top plate in the corners. Joints in the top plate must be offset by a minimum of 48 in. from joints in the double plate (see the drawing on p. 15). Overlapping provides a continuous tie in the walls.

Most of my residential-framing experience has been limited to framing 16 in. o.c. and to double top plating. When you're framing this way, there are no limitations on how close rafters, floor joists, or bottom chords of roof trusses must be to the supporting studs underneath. If you frame using a double 2x6 top plate, you can space your studs up to 24 in. o.c. and place a supporting member at any point above it (see the chart on p. 16). If it's a 2x4 frame that's 24 in. o.c. and has a double top plate, whatever is supported above it must be within 5 in. of the stud below it.

For a single top plate to be allowed, it must be spliced with a 3-in. by 6-in. by ⅛-in. steel plate at all joints (see the drawing on

Whether Load-Bearing or Not, Stud Walls Need to Stay Sturdy

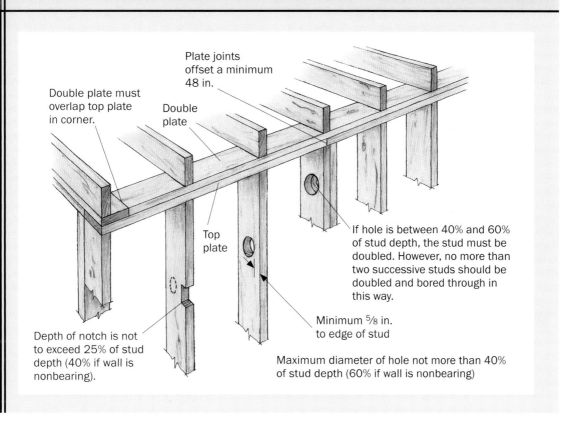

Double plate must overlap top plate in corner.

Plate joints offset a minimum 48 in.

Double plate

Top plate

If hole is between 40% and 60% of stud depth, the stud must be doubled. However, no more than two successive studs should be doubled and bored through in this way.

Minimum ⅝ in. to edge of stud

Depth of notch is not to exceed 25% of stud depth (40% if wall is nonbearing).

Maximum diameter of hole not more than 40% of stud depth (60% if wall is nonbearing)

p. 16), including corners. There also must be a minimum of three 8d nails fastening each side of the splice. Even after this, rafters, joists, or trusses must be within 1 in. of the supporting studs below them.

When Notching and Boring, Be Careful Not to Go Too Far

I once worked on a project that looked as if the plumber had just gotten his first reciprocating saw: All the notches and holes were cut in the centers of the floor joists. So much of the structural integrity of the joists was compromised by this butchering that the building inspector made us pull out the floor.

With a little planning, some of this notching and boring could have been avoided. Or we could have avoided the extra work if we had used headers to make an opening in

the joist so that no strength was lost. Now I hand out three pages of drawings to my electrical and plumbing subs that show exactly where notches and holes can and can't be cut.

Rules for notching and boring differ depending on whether the wall is a bearing wall or a nonbearing wall. Notching in a bearing-wall stud may not exceed 25 percent of the stud depth (see the drawing above). Notching in a nonbearing wall may not exceed 40 percent of the depth. If a wall was framed with 2x4s, the notch in a bearing wall may not exceed ⅞ in., as opposed to a 1⅜-in. notch in a nonbearing wall.

There also are considerable differences when it comes to boring holes in bearing and nonbearing walls. The diameter of a hole bored in a bearing wall may not exceed 40 percent of the stud depth. A bored hole in a nonbearing wall may not exceed 60 percent of the stud depth. Using a 2x4 as example, the maximum-diameter hole you

Avoiding Common Framing Errors   15

WALL CONSTRUCTION FOR ONE- OR TWO-STORY HOUSES			
Stud Size	**Stud Spacing**	**Top Plate**	**Location of Members Above**
2x4	16 in. o.c.	Single	Within 1 in. of support
2x4	16 in. o.c.	Double	Anywhere
2x4	24 in. o.c.	Single	Within 1 in. of support
2x4	24 in. o.c.	Double (2x4)	Within 5 in. of support
2x6	16 in. o.c.	Single	Within 1 in. of support
2x6	16 in. o.c.	Double	Anywhere
2x6	24 in. o.c.	Single	Within 1 in. of support
2x6	24 in. o.c.	Double (2x6 or greater)	Anywhere

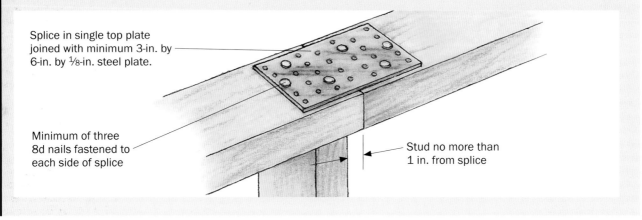

Splice in single top plate joined with minimum 3-in. by 6-in. by ⅛-in. steel plate.

Minimum of three 8d nails fastened to each side of splice

Stud no more than 1 in. from splice

may drill in a bearing wall is 1⅜ in., vs. 2⅛ in. for a nonbearing wall. In both bearing and nonbearing walls, all holes must start at a minimum of ⅝ in. from the edge. Boring and notching in the same cross section of a stud is prohibited in any case. Also, a bearing wall may have a hole whose diameter is greater than 40 percent of the stud (but less than 60 percent), if the stud is doubled and the hole doesn't carry for more than two successive studs. If the top plate is notched more than 50 percent, usually to allow for plumbing or venting, it must be reinforced with a minimum of 24-ga. steel or its equivalent.

Joists and Rafters Carry Their Own Rules

There are specific rules on notching and boring floor joists and roof rafters. No notching is allowed within the middle third of the member. Notching at the ends of floor joists, either at the top or at the bottom, may not exceed one-fourth of the depth of the joist. Notches elsewhere may not exceed one-sixth the depth of the member itself (see the drawing on p. 17). Rafters and ceiling joists may be notched at the ends but never more than one-fourth of the depth of the member. Notching on the top of the rafter or ceiling joist may not exceed one-third of the depth and may not be farther from the face of the support than the depth of the member.

Notching and Boring Floor Joists

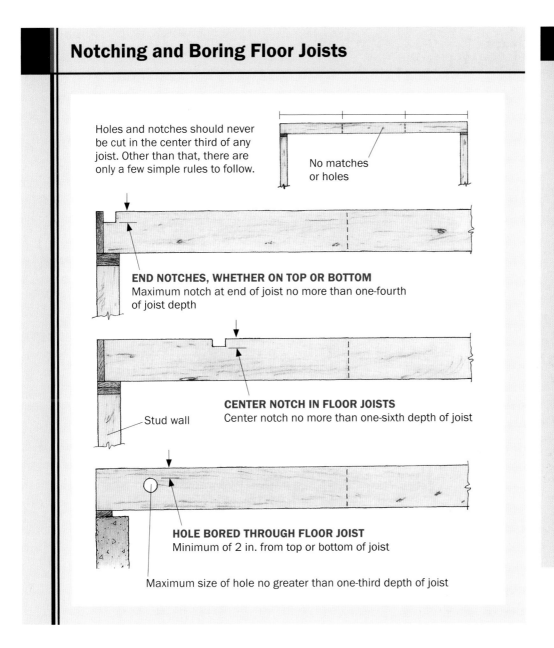

Holes and notches should never be cut in the center third of any joist. Other than that, there are only a few simple rules to follow.

No matches or holes

END NOTCHES, WHETHER ON TOP OR BOTTOM
Maximum notch at end of joist no more than one-fourth of joist depth

CENTER NOTCH IN FLOOR JOISTS
Center notch no more than one-sixth depth of joist

Stud wall

HOLE BORED THROUGH FLOOR JOIST
Minimum of 2 in. from top or bottom of joist

Maximum size of hole no greater than one-third depth of joist

Sources

Finding Code Information
Most libraries and all building-inspection offices have copies of the relevant model codes on file for public use. I recommend that you go to the source whenever you're in doubt. If you need to ask a specific question that's not clear to you in the code, you may want to take advantage of promotional trade groups.

The Southern Forest Products Association (SFPA; 504-443-4464), APA—The Engineered Wood Association (APA; 206-565-6600), and the Western Wood Products Association (WWPA; 503-224-3930) have technical representatives to answer questions. Another valuable resource for me was Redwood Kardon's *Code Check* (The Taunton Press, 2004). This clipboard-type publication puts code issues together in a simple, logical format.—R. T.

The rule for boring holes in ceiling joists and roof rafters is easy to remember because the rule is the same for both. All holes must be a minimum of 2 in. from the top and bottom. The hole may not exceed one-third of the depth of the member. Using a 2x10 for example, one-third comes out to 3$\frac{1}{16}$ in.

Rules for Engineered Wood are Evolving

The phrase engineered wood refers to combinations of lumber, plywood, metal, glue and oriented strand board, such as glue-laminated wood, laminated-veneer lumber, wood I-joists, and trusses. The Building Officials and Code Administrators Inc. (BOCA) model code says that "cuts, notches and holes bored" into engineered wood "shall be based on research and investigation" funded by companies applying for code approval of their engineered wood. This process is still under way, so until the code is clear on the use of engineered wood, we have to rely on the manufacturers' guidance. The code is clear on the use of manufactured trusses, however. They may not be cut or modified unless so specifically designed.

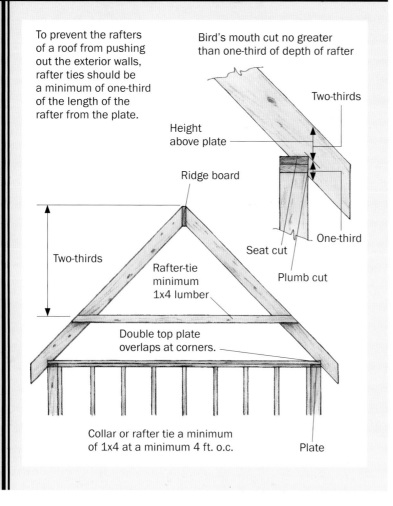

To prevent the rafters of a roof from pushing out the exterior walls, rafter ties should be a minimum of one-third of the length of the rafter from the plate.

Bird's mouth cut no greater than one-third of depth of rafter

Two-thirds

Height above plate

Ridge board

Two-thirds

One-third

Seat cut

Rafter-tie minimum 1x4 lumber

Plumb cut

Double top plate overlaps at corners.

Collar or rafter tie a minimum of 1x4 at a minimum 4 ft. o.c.

Plate

cut or bored through, nor should holes be cut in the plywood or OSB web directly over a support.

Bird's Mouths and Collar Ties Are Common Areas of Mistakes

Thinking back, I realize that most of the framing errors I've made had to do with rafter cutting. Specifically, my mistakes involved cutting bird's mouths, the notches in the ends of rafters where they rest on top plates. One reason so many errors pop up in rafter cutting is that although there are rules for notching rafters for wiring, there are no code requirements for cutting bird's mouths. I worked with one builder who believed that the seat cut of a bird's mouth should always equal the width of the wall plate on which it rested. He thought that by following this procedure, if the inside wall called for a cathedral ceiling, the drywall would go right to the edge of the wall. His method worked fine on a low-pitched roof. But with a 12-in-12 pitch roof, the bird's mouth would have to be so deep that it would eat up all but about 1 in. of a 12x12.

It was not until I read and studied Marshall Gross's book on roof framing (*Roof Framing*; Craftsman Books, 1984) that I fully understood and could solve roof-framing problems. Gross's simple solution is to leave two-thirds of the rafter material above the seat cut. In a 2x4, a 3-in-12 rafter would have 2⅜-in. height above plate (HAP) left in the member after the bird's mouth was cut. A 12-in-12 pitch roof would have a 3½-in. HAP. This measurement is made from the top of the plumb cut.

Another code violation that building inspectors cite frequently is improperly spaced or nonexistent rafter ties. Rafter or collar ties must be a minimum of 1x4 material spaced no more than 4 ft. o.c. The code only says that the rafter tie should be located as near the wall plate as is possible.

Most makers of engineered-wood materials are in sync on cutting and notching, although each offers its own set of guidelines. For instance, Willamette Industries (800-942-9927) offers an installation guide for use of its StrucJoist and StrucLam products. It cautions against any notching of its structural laminated beams. Willamette's technical department suggests that ¾-in. holes may be bored through a laminated beam as long as the holes aren't over a support, aren't too numerous or lined up, and are located in the center of the beam. If you need to bore a larger hole, technical representatives will advise you on how to proceed.

Willamette offers a hole chart for boring through its structural-wood I-joists. In no case should the flanges of a wood I-joist be

However, for a rafter tie to be effective, it must span a minimum of one-third the distance up the rafter from the top plate. David Utterback, a code expert with the Western Wood Products Association, compares rafter ties to using long-handled pruning shears. Grab the handles at the end, and you have plenty of leverage to cut off a limb. Grab the handles near the cutter, and it's much more difficult to cut a limb. So the closer the tie is to the bottom of the rafter, the greater the leverage.

Improper Nailing Can Make Itself Known the Hard Way

People don't usually notice the result of improper nailing—too few nails or nails of the wrong size—until a disaster occurs. Once again, hurricanes Hugo and Andrew showed just how many buildings were improperly nailed together. Plywood was ripped off homes. Structures collapsed before the expected limits of the construction were reached. Later investigations showed that many of these losses could have been prevented with proper nailing.

All the codes contain fastening or nailing schedules for connections of structural wood. Nevertheless, it's common to see the wrong size or type of fastener used in framing. Common mistakes include using roofing nails to hold joist hangers and using nongalvanized nails for exterior applications. Although it's not a code violation, for best results in nailing subfloors and underlayment, use ring-shank instead of smooth-shank nails.

Another factor in structural failure has to do with fastener overdrive, or overnailing, from pneumatic nailers. Even if the nailing schedule is correct, power-driven staples and nails may be misplaced (not fastened to the structural member) or overdriven because of excessive air pressure (see the drawing above right). Shear capacity also may be seriously compromised when fasteners are overdriven.

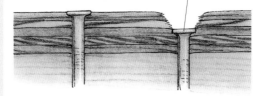

Overdriving Fasteners Reduces Sheathing Strength

A nail driven halfway into a ½-in. sheet of plywood results in the strength of a ¼-in. panel.

NAILING FOR A TYPICAL ½-IN. PANEL
The codes offer nailing schedules and nail types for a variety of panels, including panels for seismic bracing and structural and nonstructural sheathing.

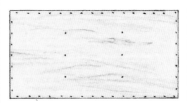

Nails 6 in. o.c. along panel edges and 12 in. o.c. along intermediate support

According to APA—The Engineered Wood Association, if more than 20 percent of the fasteners around the perimeter of a sheathing panel are overdriven, or if any are overdriven by more than ⅛ in., additional fasteners are required to maintain shear capacity. For every two fasteners that are overdriven, one additional fastener is required. If nails used in the original installation are spaced too closely to allow placement of additional nails, then approved staples must be used for the additional fasteners.

Another consideration, APA says, is the minimum nominal panel thickness required for shear design. If design shear for the construction requires a minimum $^{19}\!/_{32}$-in. nominal panel thickness and sheathing is $^{19}\!/_{32}$ in. with fasteners overdriven ⅛ in., the result is a $^{15}\!/_{32}$-in. panel.

Rick Tyrell is a builder and writer in Prosperity, South Carolina.

The Well-Framed Floor

■ BY JIM ANDERSON

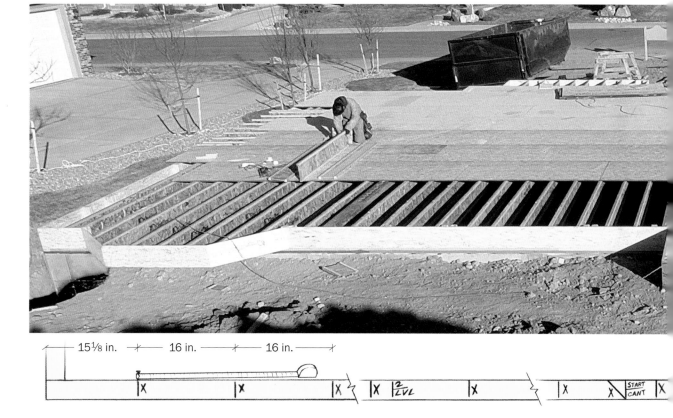

|← 15⅛ in. →|← 16 in. →|← 16 in. →|

Adjust the Layout before It's Too Late

First, lay out the mudsill for the regular 16-in. o.c. joists. Then locate additional elements, such as cantilevers, stairs, plumbing drains, and large ducts. You may need to adjust the joist spacing based on these additional elements. It's easier to make changes now than later.

Begin the layout on a long, straight section of the foundation. Place the first mark 15⅛ in. from the end for a 1¾-in. I-joist (or 15¼ in. for a solid-wood joist). From there, mark every 16 in. to the other end.

Stair openings require heavier framing (often doubled-up joists). Here, two LVLs are indicated.

Cantilevers, which require joists to extend beyond the mudsill, are labeled to indicate their angle and starting point.

Walking across a newly framed floor for the first time is a milestone in any framing project. Finally, there's something to stand on that doesn't squish beneath your boots. It's flat and strong, and because there's a floor to stand on, the rest of the project will move ahead much more quickly. But whether you're using common lumber or I-joists (see the sidebar on p. 26), it takes a well-coordinated effort to get any floor to the point where you can walk on it.

Before you start driving nails, it's important to collect as much information as possible about the locations of the joists, posts, beams, point loads, cantilevers, plumbing vents, drains, and HVAC ducts on the floor-framing plan. Whether those details come from the architect, you, or somewhere else, the floor-framing plan needs to reflect the house as it's going to be built.

Having all this information in one place allows you to overlay—in pencil—the big immovable parts of the house on top of each other. This step will catch most if not all the big mistakes that can be made early on. It's a lot easier to erase than it is to remove and replace.

Whether it's the floor of a big house or a small addition, an accurate layout and efficient techniques promote smooth installation.

If a plumbing drain, furnace flue, or duct can't be moved and it's too close to a joist, move the joist. Moving it an inch or two is better than cutting the joist and framing around the pipe.

If the last joist bay is less than 12 in., make it a full 16 in. to allow plumbers, HVAC, and electricians more room.

Bearing (or squash) blocks are installed at all point loads. But to avoid confusion with other layout marks, they aren't located and installed until after the rim joist is in place.

Transfer the Details from the Plans to the Mudsills

First, I check the joist spacing on the floor-framing plan, usually 16 in. or 19.2 in. o.c., and transfer that to the mudsills. Measuring from the end of the house (usually beginning with the longest uninterrupted run), I mark the edge of the first joist 15⅛ in. from the end for 1¾-in. I-joists (16 in. minus half the joist thickness). This places the center of the first joist at 16 in.

Then I mark 16 in. o.c. (or whatever the proper spacing is) from the first mark to the other end of the house. I do this on the front and back walls, then I check the layout marks on both ends to make sure that they are the same. If they are within ¼ in., I leave it; if not, I double-check the layout and make adjustments. I also mark the location of stairs, load-bearing members, and cantilevers on the mudsill.

Leave Room for Pipes and Ductwork

If the layout mark for the last joist is within a foot of the endwall, I move it to allow room for plumbing, electrical, or HVAC in what is often an important joist bay. I usually just measure and mark 16 in. from the edge of the mudsill back toward the center of the house.

I also make sure that none of the plumbing fixtures or flue chases lands on a joist. This is another opportunity to double-check myself. It's a lot easier to move the joist now than it is to move it later or repair damage from a determined plumber with a chainsaw. I usually allow a minimum of 12 in. between joists for furnace flues, which provides 2 in. of clearance on each side for an 8-in. furnace flue. Even though 1 in. on each side meets the building code here in Denver, I figure that where heat and wood are concerned, more room is better.

Don't Be Afraid to Hire a Crane

Many people associate cranes only with big commercial jobs, such as skyscrapers or shopping malls. But today cranes are commonly available for residential work, and anybody can hire one.

With a crane and one helper, I can set all the steel for a house and distribute stacks of presorted materials to where they're needed. This process usually takes about 1½ hours ($180 here in Denver). This easily is cheaper than paying labor to move all that material, and we get to the framing faster.

Again, I create this space either by moving the joist off the 16-in. o.c. layout or, when that isn't practical, by cutting the joist just short of the flue and supporting it with a header tied into the joists on each side of the one that's cut.

Plumbing drains and supply lines are zero-clearance items, so I can have wood right next to them. I locate the fixtures on the plan; and if a joist is on or near the centerline of the drain, I move the joist 1 in. or 2 in. in one direction or the other. If I have two fixtures close together and moving a joist away from one drain places it beneath another, I open the spacing a little more (and double the joist) so that both drains lie within a slightly oversize bay.

Prepare Material According to Where It's Needed

Wood I-joists come from the yard in a large bundle; the rim material and any LVLs usually are strapped to the top. With a helper, I move the LVLs to sawhorses for cutting to length and to install joist hangers.

We move the rim joists to the top of the sheathing or to the ground, and place stickers beneath so that we can lift them easily later. Then we square one end of all the wood I-joists with a simple jig (see the photo at right) as we take them off the pile and sort them by length and location. When I finish with the I-joists, I build any LVL headers and add joist hangers if they're needed.

After the prep work is done, I usually call in a crane to set all the steel beams that will carry the first floor and to spread all the presorted stacks of joists and LVLs to their appropriate locations. I also move the sheathing to within 3 ft. or 4 ft. of the foundation so that I don't have to carry it any farther than necessary.

After placing the steel, I make sure that the layout on the beams matches what is on the walls. I check the layout by pulling a

Square one end of each joist while sorting and stacking. Because I-joists are cut to approximate length at the lumberyard, it is easier just to square one end as you are sorting them.

string from front to back to verify that the layout marks on the front and back walls intersect the marks on the beams. I also make sure that the beams are straight and flat, and make any necessary adjustments.

Spread Joists to the Layout Marks and Roll Them Upright

With one person on each side of the foundation, we quickly position the joists on their layout marks, with the square-cut end aligned on the rim-joist line snapped along the mudsill (see the top left photo on p. 24). Then we tack each joist in place with an 8d nail to keep it in place. It's easier to set the joists to the line first and then install

Position the square ends of each joist to the chalkline (the rim-joist line) and tack them into place along their 16-in. o.c. layout lines. Later, the 8d nail will act as a hinge when the joists are stood upright.

After aligning the joists, snap a chalkline and cut them to length in place. Beware of anchor bolts lurking below when making this cut.

the rim joist later. After tacking down all the joists, we prepare to cut the other end of the joists in place. We snap a chalkline that is 1⅜ in. from the outside of the mudsill, which becomes the cutline.

Cutting the joists to their finished length is as simple as running the saw along the chalkline using the I-joist cutoff guide (see the photo at left). The scrap of wood lands in front, where it's available for use as a piece of blocking.

We position one person in the front and one in the back, and starting from one end, we stand all the joists and nail them in place (see the top photo on the facing page). The 8d nail that had held the joist in place now acts as a hinge for it. We usually can stand all of the joists for 40 lin. ft. of floor in about 10 minutes.

On each end of the I-joist and at the center beam, we put one 10d nail on each

You can do this alone, but it sure goes quicker with two. With one person at each end, stand the joists upright and put them on their layout marks. Drive one 10d nail through the flange on each side of the joist into the mudsill (or pony wall).

side of the joist through the flange into the mudsill. We keep the nails as far from the end of the joist as possible to avoid splitting the I-joist's flange. After standing the joists, we add the rim boards, cutting and nailing as we work our way around the house (see the photo at right). We put one 10d nail through the rim into the top and bottom flange of each I-joist.

Once the rim joist goes up, the last thing to do before sheathing is to add bearing blocks, also known as squash blocks (see the bottom right photo on p. 21). One person details the rim joist for bearing blocks, and another follows behind and nails them in place.

Bearing blocks are required anywhere that concentrated loads land on the joists, such as doorways or where a post supports a beam. We also put them at all inside corners because 90 percent of the time this spot is a bearing point.

The rim joist goes on after the joists are in place. The rim joists are cut and nailed to the mudsill every 8 in. with a 10d nail.

Stack Sheathing on the Floor as Soon as Possible

We snap the line for the first course of sheathing 48½ in. from the outside edge of the rim joist. It's held back a little from the rim to account for any inconsistency in the rim joist.

Before we begin nailing the sheathing, we look for joists that may have been moved from the 16-in. o.c. layout. If our plywood joints are able to avoid them, sheathing will go much faster. After deciding on a starting point, we spread construction adhesive on top of the joists. Then we lay the first row and two sheets of the second row (see the top photo on the facing page). This approach creates a little staging area where we can stack the rest of the sheathing.

We sheathe over to the steel beams in the center and add any bearing blocks and joist blocking when we get there (see the bottom photo on the facing page). Waiting until the floor is partially sheathed before installing blocks is a lot easier and safer than trying to balance on unbraced joists.

Why I Prefer I-Joists over Solid Wood

I remember the first time I saw I-joists, those long, floppy things. They seemed so flimsy and light that I thought they would have trouble holding up the sheathing, not to mention the walls that would go on top of them.

They have more than proven me wrong, however. The main advantages are that I-joists are dimensionally stable and very straight. The web (the wide middle section) of an I-joist is cut from oriented strand board, thin strands of wood oriented in the same direction and glued together. Because glue surrounds all those strands of wood,

you can expect less shrinking and swelling and very consistent joist sizes (usually within ⅟₁₆ in.).

You also can cut much larger holes into I-joists than into solid lumber; holes up to 6 in. are allowed in the center of the span of a 9½-in. I-joist. Elsewhere along the web, 1½-in. holes are provided in perforated knockouts. Holes in solid lumber can be no more than one-third the total width.

I-joists must be handled carefully; upright is best, or supported in a couple of places if carried flat. They're light, come in lengths up to 60 ft., and can span long distances as part of an engineered floor system. Best of all, they cost about the same as lumber; in the longer lengths, they actually cost less.

We cut all the blocks and spread them across the edge of the sheathing (next to the beam), starting at one end and grabbing them off the sheathing as we go. Layout marks for each joist on the plywood's edge keep the joists straight and plumb, and the spacing for blocking consistent. When we have a finished basement, we also add wall ties as we work our way across the floor, which keeps us from having to walk across unsupported joists.

When we get to a stair rough opening, we sheathe over it and brace the plywood seams. Not only is this approach safer, it also creates more usable floor space when we start framing walls. Before we stand any walls that surround the stair opening, we open it up again. If the hole is too large to sheathe over, we add a safety rail.

Lay the first row of sheathing plus two more sheets; then move the rest of the stack onto the floor. It takes about 10 minutes to move 40 sheets; it's much quicker than having to climb up and down to get every sheet.

Sheathe your way over to where blocking is needed. Do not walk across unstable joists or work from a ladder below the floor. Sheathe over to the beam, then add the joist blocking.

Cut the sheathing in place. Run sheets long at the ends, snap a chalkline, and cut off the excess. This process is faster and turns out a better floor than cutting each piece to fit.

Lay as Many Full Sheets as Possible

As we sheathe, we lay as many full sheets as possible (making the fewest number of cuts). I've found that running the sheets long at the ends and cutting to a chalkline snapped along the rim (see the photo above) turns out a better product than measuring and cutting the pieces to fit individually.

I pull the chalkline in an extra ⅛ in. from the outside of the rim; this eliminates ever having to cut the rim line again. One person starts at a corner of the house and snaps all the rim lines; the other follows behind with the saw. The rim joist is first straightened and then nailed to the sheathing every 6 in.

Jim Anderson is a framing contractor in Littleton, Colorado.

Careful Layout for Perfect Walls

■ BY JOHN SPIER

Framing walls is one of the most fun parts of building a house. It's fast, safe, and easy, and at the end of the day, it's satisfying to admire the progress you've made. Before cranking up your compressor and nailers, though, you need to think through what you're going to do. You need to locate every wall precisely on the subfloor, along with every framing member in those walls.

Layout Starts in the Office

For one of our typical houses, layout and framing for interior and exterior walls start in the office a few days before my crew and I are ready to pick up the first 2x6. First, I review the plans carefully and make sure that all the necessary information is there.

I need the locations and dimensions of all the rough openings, not only for doors and windows but also for things such as fireplaces, medicine cabinets, built-ins, dumbwaiters, and the like. I also make sure the plans have the structural information I need for layout, such as shear-wall and bearing-wall details and column sizes.

At the site, one of Spier's many corollaries to Murphy's Law is that errors never cancel each other out; they always multiply. If the floor is anything but straight, level, flat, and square, the walls are going to go downhill (or uphill) from there. So before you get to layout, do whatever it takes to get a good floor, especially the first: Mud the sills, shim the rims, rip the joists. Sweep off the subflooring, and avoid the temptation to have a pile of material delivered onto it.

Snap Chalklines for the Longest Exterior Walls First

I've learned over the years that it's best to snap the plate lines for the entire floor plan before building any of it. Problems you didn't catch on the prints often jump out when you start snapping lines.

When framing floors, I take great care to set the mudsills flush, square, and in their exact locations. Because the edges of floor framing and subflooring are not always perfect, though, I use a level to plumb up from the mudsills and establish the plate lines, measuring in the stock thickness from the level (see the top left drawing on p. 32). I generally start with the longest exterior walls and the largest rectangle in the plan. When I have the ends of the longest wall located, I snap a line through the marks.

Everything from **rafters** to kitchen cabinets fits better when you get the walls square and the studs in the right places.

Once I've established the line for the first wall, I move to the parallel wall on the opposite side of the house. I measure across the floor from the first line to the opposite mudsill (again using a level to plumb up from the mudsill to the floor height) at both ends; if the lengths differ slightly, I use the larger measurement. I snap through these points, which gives me two parallel lines representing the long sides of the largest rectangle (see the drawings at the bottom of p. 32). It's okay if the plates overhang the floor framing by a bit, but I watch for areas that might need to be shimmed or padded—for instance, where a deck ledger needs to be attached to the house.

Establish the Right Angles

I locate three corners by measuring in from the mudsills. The fourth corner I locate by duplicating the measurement between the first and second because I need sides of equal lengths to create a rectangle. I check this rectangle for square by measuring both its diagonals (see the right photo on p. 32). If I've done everything right so far, the diagonal measurements should be very close, perhaps within ¼ in. I shift two corners slightly if I need to, making sure to keep the lengths of the sides exact until the diagonals are equal. Now, perpendicular lines are snapped through the corners, completing the rectangle.

Because I started arbitrarily with one long wall, I may find now that the rectangle, while being perfectly square, is slightly askew from the foundation and floor. Also, some of the complicated foundations that I work on can have wings or jogs that are slightly off. If I can make everything fit better by rotating the rectangle slightly, I take the time to do it now.

Chalkline tip. To snap chalklines for short walls, hook one end of the line to your boot and stretch the other end out to the mark. Rotating your foot slightly aligns the boot end, and you're ready to snap.

Smaller Rectangles Complete the Wall Layout

With the largest part of the plan established, I lay out and snap whatever bays, wings, and jogs remain for the exterior walls. I use a series of overlapping and adjacent rectangles, which I can square by keeping them parallel to the lines of the original rectangle. I again check the right angles by measuring the diagonals.

Often, the plan calls for an angled component such as a bay. If these components are at 45 degrees, I lay them out from right angles by forming and diagonally bisecting a square. For angles that are not 45 degrees, I either can trust the architect's measurements on the plans or I can use geometry and a calculator. The latter method is more likely to be accurate.

TIP

When reviewing the plans and laying out walls, watch for elements of the design that need to stay symmetrical and make sure symmetrical elements are aligned at the first layout stage.

For the most precise wall layout, plot a series of rectangles that includes every wall. The larger the rectangle, the more accurate the wall position. Begin with the longest walls, and lay out the largest rectangle using diagonal measurements (photo below). Working off established lines and square corners, work down to the smallest rectangle.

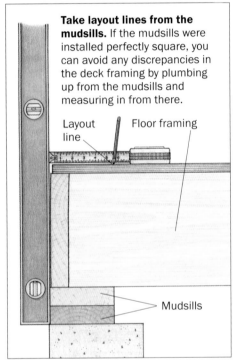

Take layout lines from the mudsills. If the mudsills were installed perfectly square, you can avoid any discrepancies in the deck framing by plumbing up from the mudsills and measuring in from there.

Layout line

Floor framing

Mudsills

Equal diagonal measurements mean a square layout. After snapping chalklines for the longest parallel walls, the author takes corner-to-corner measurements to make sure the corners are square for a perfect rectangle.

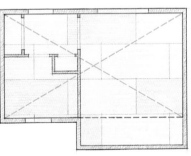

1. Starting with the longest walls, measure and square the largest rectangle.

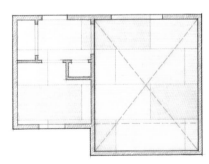

2. Working off those lines, plot the rectangle that includes the jog in the wall.

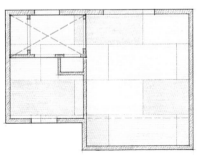

3. Now measure off the outside and form a rectangle for the longest interior wall.

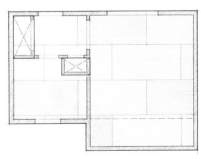

4. Last, form rectangles for the remaining interior walls.

Two Quick Layout Tips

Laying Out Multiples

For things such as short closet wall plates, line them up and draw two walls' worth of layout lines at once (below left).

Copy the Layout from the Plate

To mark the cripple layout on the rough windowsill, just line it up on the plate and copy the layout (below right).

When all the exterior walls have been laid out, I turn my attention to the interior walls. Again, I start with the longest walls and work to the smallest, snapping lines parallel and square to the established lines of the exterior walls. I snap only one side of each plate, but I mark the floor with an X here and there to avoid confusion about where the walls will land. I also write notes on the floor to indicate doors, rooms, fixtures, bearing walls, and other critical information.

It's a rare architect who dimensions a plan to a fraction of an inch with no discrepancies, and an even rarer builder who achieves that accuracy. So first, I lay out critical areas such as hallways, stairwells, chimneys, and tub or shower units, and then I fudge the rest if I need to.

One last critical issue when reviewing the plans and laying out walls is watching for elements of the design that need to stay symmetrical. If the foundation contractor made one wing a bit wider than another, you don't want to build all three floors before realizing that the ridgelines of the two wings needed to match up. Make sure symmetrical elements are aligned at the first layout stage.

Make Plates from the Straightest Lumber

While I snap the walls, the crew is busy cutting and preparing material from the piles of stock. I have them set aside a pile of the straightest lumber. With the chalklines all snapped and with this material in hand, I start cutting and laying out the plates (top and bottom members) for the exterior walls (see the photo on the facing page). In this step of layout, we set the plates side by side on the layout line, and every wall-framing member is located and labeled. With this information, we assemble the walls on the

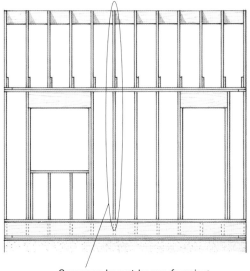

Common layout keeps framing members in line

Pulling the layout from the same point for every level of the house lines up the joists, studs, and rafters for a stronger house that's easier to finish.

Stud layout is always taken from the same two walls. One crew member holds the tape at wall offset while the other marks the stud position (top). Even when there is a break in the wall, the layout is pulled from the same place to keep all the framing aligned (above).

Plates on deck. When wall positions are laid out, cut all the top and bottom plates (the long horizontal members) for the exterior walls, and place them on their layout lines.

floor, then raise them into place. I often call out measurements and have someone cut and hand up the material to keep mud, snow, and sawdust off the floor during this crucial phase.

As a rule, we plate the longest exterior walls to the corners of the house, and the shorter walls inside them. This approach sometimes needs to be modified—for instance, to accommodate structural columns, hold-down bolts, or openings adjacent to corners. Sometimes an obstruction or a previously raised wall dictates which wall can be built and raised first. The goal here is to build and raise as many walls as possible in their exact positions, especially the heavier ones. Moving walls after they're raised is extra work and no fun.

Before starting any framing, I established a common-stud layout for the entire structure based on two long perpendicular walls from which layout for the rest of the house framing can be measured (see the drawing on the facing page). This common layout keeps joists, studs, cripples, and rafters throughout the house vertically aligned from the foundation to the ridge, which makes for a strong, straight, and easily finished structure. We use this common layout to locate butt joints between pairs of plates because code and common sense dictate that these joints land on a stud or a header.

As my crew and I measure and cut the pairs of wall plates, we lay them on edge along their layout lines, sometimes tacking them together with just a few 8d nails to keep the plates held together and in place.

Window, Door, and Stud Layout at Last

When all the exterior plates are in place, it's finally time to lay out the actual framing members. I always start with the rough openings for windows and doors. Most plans specify these openings as being a measured distance from the building corner to the center of the opening, which works fine. You can allow for the sheathing thickness or not, but once you choose, be consistent, especially if openings such as windows have to align vertically from floor to floor. Obviously, if an opening such as a bay window or a front door is to be centered on a wall, center it using the actual dimensions of the building, which may differ slightly from the plan.

Framer's Shorthand: What Those Little Marks Mean

When the framing members are marked, a full-length stud is indicated by an X. A trimmer or jack stud is a T or J, and a C or X indicates a cripple (a short framing member below a sill or above a nonstructural header). Other framing, such as partition posts and corner posts, are labeled along with any special framing instructions.

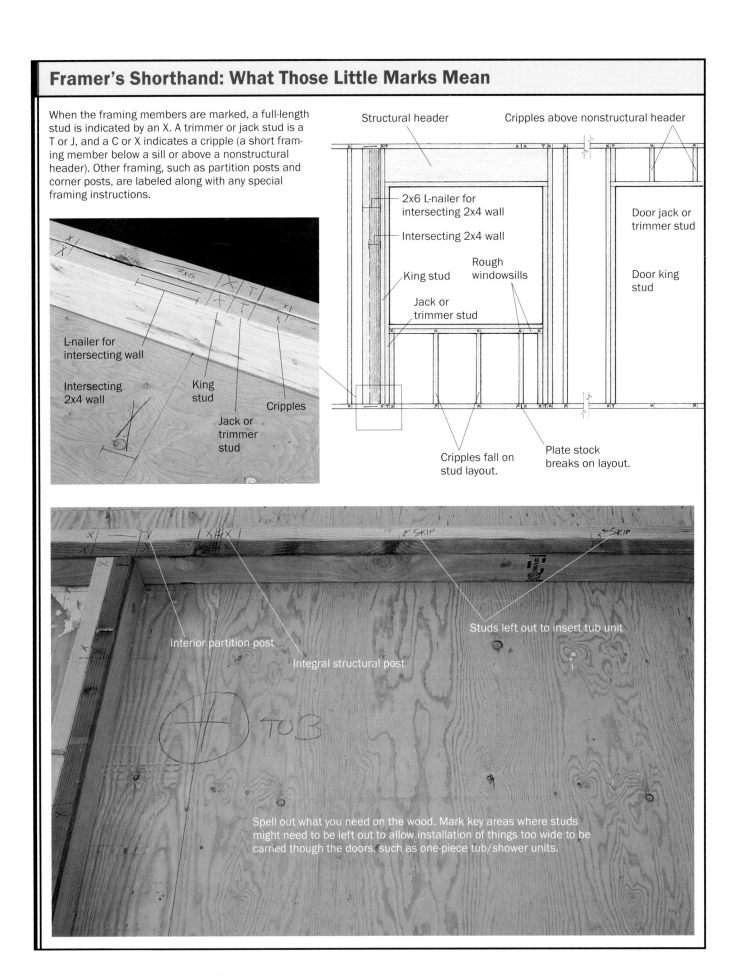

Structural header

Cripples above nonstructural header

2x6 L-nailer for intersecting 2x4 wall

Intersecting 2x4 wall

King stud

Rough windowsills

Jack or trimmer stud

Door jack or trimmer stud

Door king stud

Cripples fall on stud layout.

Plate stock breaks on layout.

L-nailer for intersecting wall

Intersecting 2x4 wall

King stud

Jack or trimmer stud

Cripples

Interior partition post

Integral structural post

Studs left out to insert tub unit

TUB

Spell out what you need on the wood. Mark key areas where studs might need to be left out to allow installation of things too wide to be carried though the doors, such as one-piece tub/shower units.

Rough openings are a subject worthy of their own chapter, but in a nutshell, I measure half the width of the opening in both directions from the center mark. I then use a triangular square to mark the locations of the edges of trimmers and king studs, still working from the inside of the opening out. Various other marks, such as Xs or Ts, identify the specific members and their positions (see the drawing on the facing page).

Next, I mark where any interior-wall partitions intersect the exterior wall. At this point, I just mark and label the locations; I decide how to frame for them later. I also locate and mark any columns, posts, or nailers that need to go in the wall. I lay out any studs that have to go in specific locations for shelf cleats, brackets, medicine cabinets, shower valves, cabinetry, ductwork, and anything else I can think of.

Doing this layout now is much easier than adding or moving studs later.

Finally, I lay out the common studs on the plates. Studs are commonly spaced either 16 in. or 24 in. o.c. to accommodate standard building products. By doing the common-stud layout last, I often can save lumber by using a common stud as part of a partition nailer. I almost never skip a stud because it's close to another framing member, which, I've learned the hard way, almost always causes more work than it saves. I occasionally shift stud or nailer locations to eliminate small gaps and unnecessary pieces. I keep the plywood layout in mind here, though, so that I can use full sheets of sheathing as much as possible.

Inside Walls Go More Quickly

Once the exterior walls are built and standing, I cut the interior-wall plates and set them in place. Where two walls meet, I decide which one will run long to form the corner so that the walls can be built and raised without being moved. Also, facing a corner in a particular direction

often provides better backing for interior finishes, such as handrails or cabinetry, and sometimes is necessary to accommodate such things as doorways or multiple-gang switches.

When the plates are cut and set in place, I do the stud layout. Just as with the exterior walls, I do the openings first, then nailers and specific stud and column locations. Next, I mark the locations of intersecting walls and finally overlay the common-stud layout on the plates.

Where Walls Come Together

Where one wall meets the middle of another, I use a partition post if the situation dictates it, but more often, I opt for an L-nailer. To make an L-nailer, I use a wider stud on the flat next to a common stud whenever possible. It's faster and easier; it accommodates more insulation; and it saves the subs from drilling through those extra studs and nails. If I use U-shaped partition posts (a stud or blocks on the flat flanked by two other studs) in an exterior wall, I need to make sure to fill the void created by the partition post with insulation before the sheathing goes on.

With the interior plates all there, we can nail in the studs, raising walls as we go. I mark key areas where studs should be crowned or specially selected, such as areas with long runs of cabinetry, and also studs that might need to be left out to allow installation of things too wide to be carried though the doors. I also nail double top plates to as many walls as possible if they don't interfere with the lifting process.

John Spier and his wife, Kerri, own Spier Construction, a custom building company on Block Island, Rhode Island. He is the author of Building with Engineered Lumber (Taunton Press, 2006).

All About Headers

■ BY CLAYTON DEKORNE

Like many carpenters in the Northeast, I was taught to frame window and door headers by creating a plywood-and-lumber sandwich, held together with generous globs of construction adhesive and the tight rows of nails that only a nail gun could deliver. Years later, I learned my energetic efforts to build a better header were an exceptional waste of time and resources. Neither the plywood nor the adhesive contributed much strength, only thickness, and this perfect thickness helped only to conduct heat out of the walls during the severe winters common to the region.

At the same time I was laying up lumber sandwiches, young production framers on the West Coast were framing headers efficiently using single-piece 4x12s. They needed only to be chopped to length and filled the wall space above openings, eliminating the need for maddeningly short cripple studs between the top of the header and the wall plate. Nowadays, however, such massive materials are relatively scarce and remarkably expensive, even on the West Coast. So although solid-stock headers certainly save labor, they no longer provide an economical alternative.

With these experiences in mind, I set out to discover some practical alternatives, surveying a number of expert framers in different regions of the country. Header framing varies widely from builder to builder and from region to region. Even when factors such as wall thickness and load conditions are made equal, building traditions and individual preferences make for a wide range of header configurations. The examples shown here are just a few of the options possible when you mix and match features, notch cripple studs, and sift in engineered materials. But they aptly demonstrate a number of practical considerations that must be kept in mind when framing a good header.

Big Headers Need More Studs

A header transfers loads from the roof and floors above to the foundation below by way of jack studs (see the drawing on p. 40). This means the header not only must be deep enough (depth refers to the height of a beam: 2x10s are deeper than 2x6s) for a given span to resist bending under load, but also must be supported by jack studs on each end that are part of a load path that continues to the foundation.

The header you choose affects not just strength, but also cost, energy efficiency, and drywall cracking.

Like a Bridge over a Ravine, a Header Spans a Window or Door

Headers are short beams that typically carry roof and floor loads to the sides of openings for doors or windows. Jack studs take over from there, carrying the load to the framing below and eventually to the foundation. That's called the load path, and it must be continuous. The International Residential Code (IRC, the most common code nationwide) has a lot to say about headers, including the tables you need to determine the size header required for most situations. If it's not in the IRC, you need an engineer.

Double 2x4

WHERE YOU DON'T NEED A STRUCTURAL HEADER

In a gable-end wall that doesn't support a load-bearing ridge, or in an interior nonbearing partition, headers up to 8 ft. in length can be built with the same material as the wall studs. Inside, a single 2x is often sufficient, although a double 2x is helpful for securing wide trim. In a gable-end wall, however, a double 2x is needed to help resist the bending loads exerted by the wind.

Load path must be continuous from roof to foundation.

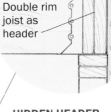

Double rim joist as header

HIDDEN HEADER

Here's a wood-saving trick. By simply adding an additional member to it, the rim joist above an opening can serve as a header. Two caveats: You'll need to use joist hangers to transfer the load to the header. And if a standard header of this size requires double jack studs, so does a double rim joist.

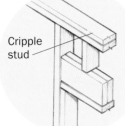

Cripple stud

CRIPPLE STUDS FILL THE VOID

If the header doesn't fill the space all the way to the top plate, cripple studs are used to carry the load from the rafters, joists, or trusses above to the header below. For nonbearing headers, the IRC requires no cripples if the distance to the top plate is less than 24 in.

King stud

Header

Jack stud

KINGS HIGH, JACKS LOW

King studs are the same height as the wall studs, running plate to plate. Nails driven through them into the header's end grain stabilize the header. Jack studs are shorter and fit below the header to carry loads downward. Because longer-spanning headers usually carry greater loads, you may need an extra jack under both ends of big headers. Check your building code.

Installation Guidelines

Typically, header height is established by the door height, and window headers are set at this same height. In homes having 8-ft. ceilings, a header composed of 2x12s or of 2x10s with a flat 2x4 or 2x6 nailer on the bottom accommodates standard 6-ft. 8-in. doors, as shown in the drawing at right.

In a custom home with cathedral ceilings and tall walls, however, header heights can vary widely. And if the doors are a nonstandard height, you'll need to figure out the header height. Finding the height of the bottom of the headers above the subfloor is a matter of adding up the door height, the thickness of the finished-floor materials, and 2½ in. (to allow space for the head jamb and airspace below the door). There are exceptions. Pocket doors typically require a rough opening at least 2 in. higher than a standard door. Windows may include arches or transoms, which affect the rough opening's height.

To find the header length for windows, add 3 in. to the manufacturer's rough-opening dimension if there is to be one jack stud on each side, or 6 in. if two jacks are called for. For doors with single jack studs, add 5½ in. to the door width to allow for jack studs, door jambs, and shim space. If double jacks are needed, then the header should be 8½ in. longer than the door width.

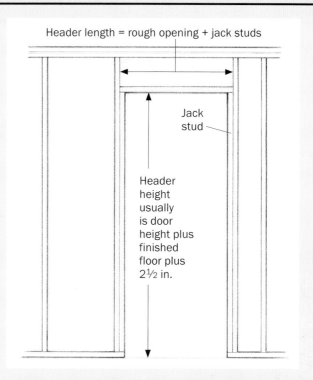

Header length = rough opening + jack studs

Jack stud

Header height usually is door height plus finished floor plus 2½ in.

These guidelines follow one fundamental rule of framing rough openings: Know your windows and doors. If you don't have the window or door on site, at the very least check the manufacturer's catalog to verify the rough-opening dimensions. Don't rely on the plans alone, and when in doubt, call the manufacturer.

The International Residential Code (IRC) specifies not only header size but also the number of jack studs for most common situations. While most windows and doors require just one jack stud at each end, long spans or extreme loads may call for two or more jack studs to increase the area bearing the load. If the loads on any header are concentrated over too small an area, the wood fibers at the ends of the header can be crushed. This can cause the header to drop, which in turn can crack drywall or, particularly with patio doors and casement windows, cause the door or window to jam.

Header hangers, such as the Simpson Strong Tie® HH Series (see the photo on p. 43), can be used to eliminate jack studs altogether. I've used them in some remodeling situations when I needed to squeeze a patio door or a wide window into an existing wall that didn't have quite enough space for double jack studs. One jack and a Simpson HH Series hanger did the trick.

TIP

While most carpenters tend to think that more nails are a sign of good workmanship, headers need to be nailed with only one 10d common nail every 16 in. along each edge.

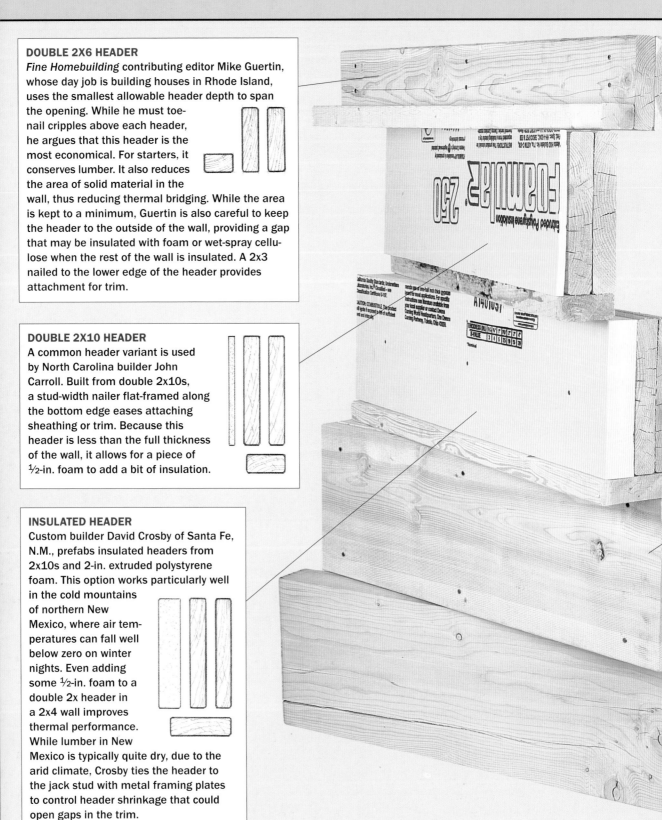

DOUBLE 2X6 HEADER

Fine Homebuilding contributing editor Mike Guertin, whose day job is building houses in Rhode Island, uses the smallest allowable header depth to span the opening. While he must toe-nail cripples above each header, he argues that this header is the most economical. For starters, it conserves lumber. It also reduces the area of solid material in the wall, thus reducing thermal bridging. While the area is kept to a minimum, Guertin is also careful to keep the header to the outside of the wall, providing a gap that may be insulated with foam or wet-spray cellulose when the rest of the wall is insulated. A 2x3 nailed to the lower edge of the header provides attachment for trim.

DOUBLE 2X10 HEADER

A common header variant is used by North Carolina builder John Carroll. Built from double 2x10s, a stud-width nailer flat-framed along the bottom edge eases attaching sheathing or trim. Because this header is less than the full thickness of the wall, it allows for a piece of ½-in. foam to add a bit of insulation.

INSULATED HEADER

Custom builder David Crosby of Santa Fe, N.M., prefabs insulated headers from 2x10s and 2-in. extruded polystyrene foam. This option works particularly well in the cold mountains of northern New Mexico, where air temperatures can fall well below zero on winter nights. Even adding some ½-in. foam to a double 2x header in a 2x4 wall improves thermal performance. While lumber in New Mexico is typically quite dry, due to the arid climate, Crosby ties the header to the jack stud with metal framing plates to control header shrinkage that could open gaps in the trim.

How Big a Header Do You Need?

Unless you're an engineer, the easiest way to size headers built with dimensional lumber is to check span charts, such as those in the IRC. The old rule of thumb is that headers made of double 2x stock can span safely in feet half their depth in inches. So by this rule, a double 2x12 can span 6 ft.

However, header spans vary not only with size, but also with lumber grade and species, with the width of the house, with your area's snow load, and with the number of floors to be supported. Consequently, the IRC provides 24 scenarios in which that double 2x12 header can span a range from 5 ft. 2 in. to 9 ft. 9 in. Check the code.

The Trouble with Cripples

Header size often is based on factors other than strength requirements. Many framers purposely oversize headers to avoid filling the space between the header and the double top plate with short studs (cripple studs, or cripples). In a nominal 8-ft.-tall wall, a

BUILT-UP PLYWOOD AND LUMBER

In this header sandwich, plywood adds only thickness so that the header will fit flush to each face of the wall. There is little strength added, even if the header is spiked together with construction adhesive between each layer. Construction adhesive adds nothing to the strength of a beam. Before assembling this (or any other header), crown the lumber, marking it clearly with a lumber crayon, and keep the crown up. Rip plywood ½ in. narrower than the lumber to prevent the pieces from hanging over the edges, especially if the lumber has a crown.

SOLID-STOCK HEADER

Once standard fare for West Coast production framers, a solid header made with a single 4x12 tucks tight under the top plates in a wall, eliminating the need for short cripple studs. Although this option saves substantial labor, the availability of full-dimension lumber is limited mainly to the West Coast. Even there, solid-stock headers are expensive and may not be cost-effective unless the opening requires the load-bearing capacity of such large-dimension stock.

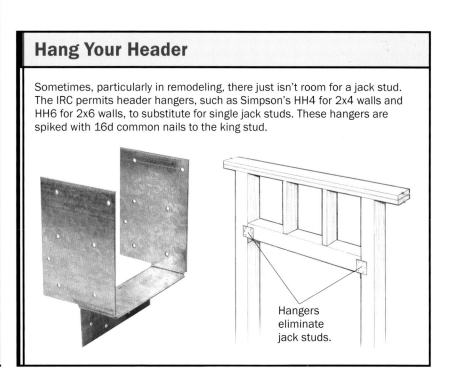

Hang Your Header

Sometimes, particularly in remodeling, there just isn't room for a jack stud. The IRC permits header hangers, such as Simpson's HH4 for 2x4 walls and HH6 for 2x6 walls, to substitute for single jack studs. These hangers are spiked with 16d common nails to the king stud.

Hangers eliminate jack studs.

Sources

APA—The Engineered Wood Association
Nailed Structural-Use Panel and Lumber Beams
Available online
www.apawood.org

Simpson
4637 Chabot Dr., #200
Pleasanton, CA 94588
800-999-5099

Superior Wood
1301 Garfield Ave.
Superior, WI 54880
800-375-9992
www.swi-joist.com

TrusJoist MacMillan's Parallam
www.trusjoists.com

typical cripple stud measures 6 in. to 7 in. Such short studs are ungainly and are prone to splitting when they are nailed in place. Yet a double 2x12 header can be tucked beneath the double top plate, filling this miserable space and creating a proper opening for common 6-ft. 8-in. doors. Alternatively, builder John Carroll relies on a double 2x10 header with a 2x6 nailed flat along the bottom edge, which provides nailing for the head trim in a 2x6 wall.

However, such deep headers are oversize and add considerable cost, not to mention waste wood. Most window and door openings are only 3 ft. or so and might only require 2x6 headers. But perhaps the biggest drawback of wide lumber is that there's more of it to shrink. Framing lumber may have a moisture content of 19 percent. Once the heat is turned on, lumber typically dries to a moisture content of 9 percent to 11 percent, shrinking nominal 2x10s and 2x12s as much as ¼ in. across the grain. On the other hand, 2x6s might shrink only half that.

Shrinkage reduces the depth (or height) of the header; because the header is nailed firmly to the double top plate, a gap usually opens above the jack studs. As the header shrinks, it tends to pull up the head trim, which has been nailed to it, opening unsightly gaps in the casing and cracking any drywall seam spanning the header. The gap above the jack stud now means the header isn't supporting any load—until the first wet snowfall or heavy winds bring a crushing load to bear on the wall and push the gap closed, causing the top plates to sag, which can crack the drywall in the story above.

Shrinkage can be reduced using drier lumber, preferably at about 12 percent. However, lumber this dry may be difficult to find unless you can condition it yourself. As an alternative for spanning a large opening, consider using engineered materials (see photo on facing page). Laminated-veneer lumber (LVL) or parallel-strand lumber (PSL) shrinks much less than ordinary lumber.

If wide dimensional lumber is unavoidable, structural engineer Steve Smulski sug-

gests that cracking can be minimized by not fastening the drywall to the header. This way, the header moves independently of the drywall, which then is less likely to crack. To prevent trim from moving as the header shrinks, attach the top piece of trim to the drywall only, using a minimal number of short, light-gauge finish nails and a bead of adhesive caulk.

Avoiding Condensation

In cold climates, uninsulated headers can create a thermal bridge. According to Smulski, the uninsulated header makes the wall section above windows and doors significantly colder than the rest of the wall (the same is true of solid-frame corners). When the difference between the inside and outside air temperatures is extreme, condensation may collect on these cold surfaces, and in the worst cases, mold and mildew may begin to grow.

To avoid condensation, it's important that any uninsulated header doesn't contact both the sheathing and the drywall. Unless you're building 2x4 exterior walls using full-thickness headers such as solid lumber or ones built out to 3½ in. with plywood, avoiding this situation is simple. Keep the header flush to the outside of the framing so that it contacts the sheathing. Because most other types of headers are narrower than the studs, there will be some airspace between the header and the drywall, which makes a dandy thermal break. In cold climates, a 2x10 insulated header, like the one used by David Crosby of Santa Fe, N.M., works well (see the photo on p. 42). Another option that avoids solid lumber is a manufactured insulated I-beam header (see the photo on the facing page).

Clayton DeKorne is a carpenter and writer in Burlington, Vermont. He is the author of Trim Carpentry and Built-Ins *(The Taunton Press, Inc., 2002). His book,* Finish Carpentry *will be published by The Taunton Press in 2008.*

Engineered Wood: Costs More, Does More

STORE-BOUGHT INSULATED HEADERS

Essentially a double-webbed I-joist with a chunk of rigid foam wedged in the middle, these engineered SW-II headers from Superior Wood Systems® offer insulation, strength, and light weight. You may have a hard time finding them locally, though, because they're new enough that distribution varies regionally. Price varies as well, depending on freight costs and markup. Hammond Lumber in Bangor, Maine, sells 14-ft. long, 5½-in. by 11¼-in. SW-II headers for about $90.

PARALLEL-STRAND LUMBER

Parallel-strand lumber, such as TrusJoist® Mac-Millan's Paral-lam, is available as stud-width stock. Perform-ing much like the LVL above it, parallel-strand header stock is pricier than solid sawn lumber but 1½ times as stiff and 3 times as strong.

LAMINATED-VENEER LUMBER

Engineered lumber, shown in this header made from two pieces of 1¾-in. by 16-in. LVL (laminated-veneer lumber), offers some advantages over sawn lumber. While it's more expensive for smaller headers, engineered lumber is available in depths that can span distances sawn lumber simply isn't up to. And it's typically more stable, resulting in fewer drywall cracks.

STRUCTURAL BOX BEAM

A box-beam header is a viable way to site-build long-span headers. A technical bulletin, Nailed Structural-Use Panel and Lumber Beams, outlines the design and fabrication of these stud and plywood beams. Because they end up being thicker than the studs, these plywood beams are better suited for long-span headers in an unfinished garage, where the exact thickness is a slight concern. For a 2x6 wall, though, you can make a box beam using 2x4 blocking and nominal ¾-in. structural plywood. A ½-in. furring strip brings such headers to the full wall thickness. And they can be stuffed with insulation.

Mudsills: Where the Framing Meets the Foundation

■ BY JIM ANDERSON

Framing a traditional house begins at the mudsill; it's the first piece of lumber that is attached to the foundation. If you build on a foundation that's out of square or level (and they're common), correcting the problem at the mudsill stage will make for a lot less trouble later. The first step in installing mudsills is determining if the foundation is square.

Checking for a Square Foundation

While my helper sweeps off the foundation and checks to make sure the anchor bolts are plumb, I look over the plans for the foundation's largest rectangle. It will provide an ongoing reference for establishing bump-outs (areas outside the large rectangle) and recesses (areas within the large rectangle) that are square to the house and to each other. The

drawing on pp. 48–49 shows how to square the large rectangle or to create a large 3-4-5 triangle if a rectangle can't be found.

After squaring and marking its corners, two of us snap chalklines for the large rectangle and any bump-outs or recesses, while a third person spreads pressure-treated 2x4s (or 2x6s if requested) around the foundation to serve as mudsills. Working as a team with a systematic approach is really important on these projects. We begin at the front corner and run the material along the chalkline from end to end, and then do the same in the rear. We fill in the sidewalls last.

As we work our way around the foundation, we mark the bolt locations by standing the plates on edge and outlining the bolts on the sill plate (see the left photo on p. 50). When we have to join two mudsills, we cut the first plate within 12 in. of a bolt and add an expansion bolt for the adjacent plate.

Even if the foundation is not square or level, getting the mudsills right will get you back on track.

Square Mudsills Depend on a Square Layout

Before you can install the mudsills, you have to make sure that the foundation is square. And occasionally, foundations are a little out of square. To get the framing off to a good start, you have to snap a series of square layout lines for the mudsills. Locating and snapping lines on the foundation's largest rectangle provides a square reference for the remaining areas lying inside or outside the large rectangle. If a large rectangle can't be found, a large right triangle (see "Finding Square with Triangles" on p. 50) will work.

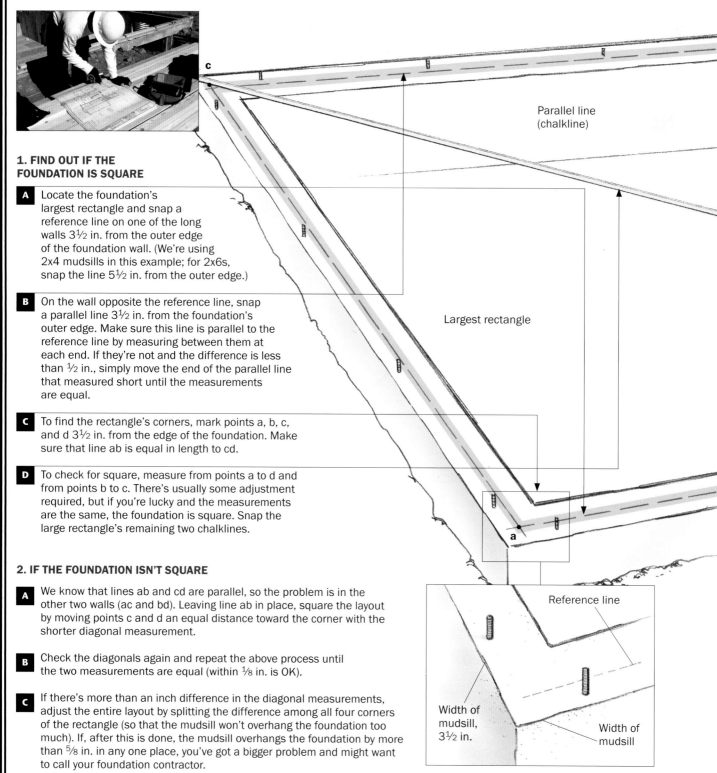

Parallel line
(chalkline)

Largest rectangle

Reference line

Width of mudsill, 3½ in.

Width of mudsill

1. FIND OUT IF THE FOUNDATION IS SQUARE

A Locate the foundation's largest rectangle and snap a reference line on one of the long walls 3½ in. from the outer edge of the foundation wall. (We're using 2x4 mudsills in this example; for 2x6s, snap the line 5½ in. from the outer edge.)

B On the wall opposite the reference line, snap a parallel line 3½ in. from the foundation's outer edge. Make sure this line is parallel to the reference line by measuring between them at each end. If they're not and the difference is less than ½ in., simply move the end of the parallel line that measured short until the measurements are equal.

C To find the rectangle's corners, mark points a, b, c, and d 3½ in. from the edge of the foundation. Make sure that line ab is equal in length to cd.

D To check for square, measure from points a to d and from points b to c. There's usually some adjustment required, but if you're lucky and the measurements are the same, the foundation is square. Snap the large rectangle's remaining two chalklines.

2. IF THE FOUNDATION ISN'T SQUARE

A We know that lines ab and cd are parallel, so the problem is in the other two walls (ac and bd). Leaving line ab in place, square the layout by moving points c and d an equal distance toward the corner with the shorter diagonal measurement.

B Check the diagonals again and repeat the above process until the two measurements are equal (within ⅛ in. is OK).

C If there's more than an inch difference in the diagonal measurements, adjust the entire layout by splitting the difference among all four corners of the rectangle (so that the mudsill won't overhang the foundation too much). If, after this is done, the mudsill overhangs the foundation by more than ⅝ in. in any one place, you've got a bigger problem and might want to call your foundation contractor.

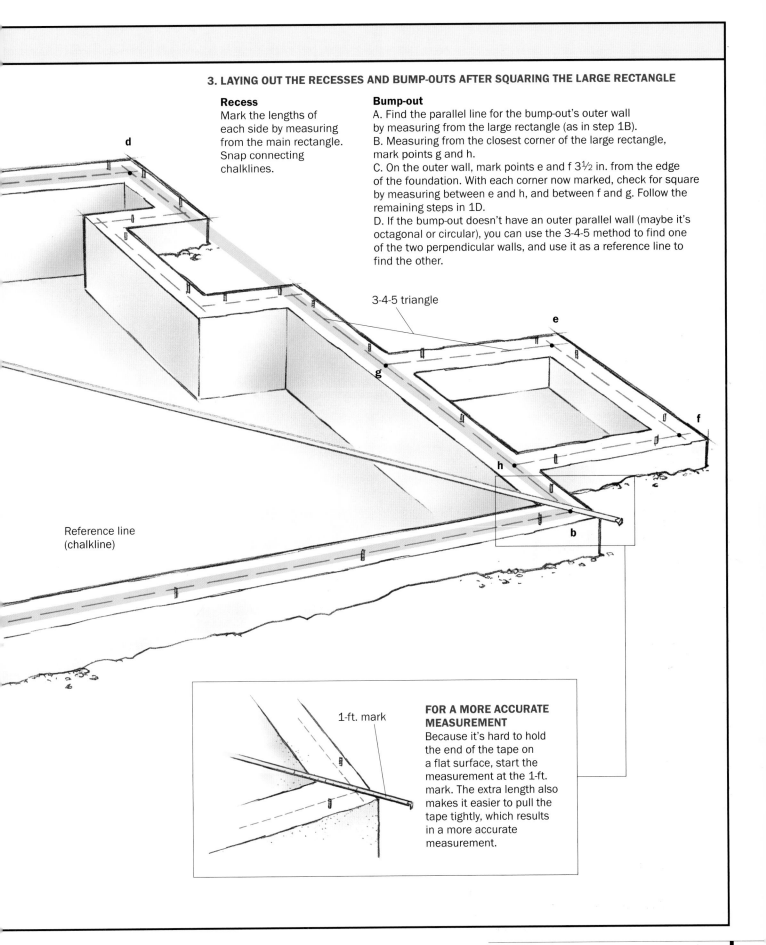

3. LAYING OUT THE RECESSES AND BUMP-OUTS AFTER SQUARING THE LARGE RECTANGLE

Recess
Mark the lengths of
each side by measuring
from the main rectangle.
Snap connecting
chalklines.

Bump-out
A. Find the parallel line for the bump-out's outer wall
by measuring from the large rectangle (as in step 1B).
B. Measuring from the closest corner of the large rectangle,
mark points g and h.
C. On the outer wall, mark points e and f 3½ in. from the edge
of the foundation. With each corner now marked, check for square
by measuring between e and h, and between f and g. Follow the
remaining steps in 1D.
D. If the bump-out doesn't have an outer parallel wall (maybe it's
octagonal or circular), you can use the 3-4-5 method to find one
of the two perpendicular walls, and use it as a reference line to
find the other.

d

3-4-5 triangle

e

g

f

h

b

Reference line
(chalkline)

1-ft. mark

**FOR A MORE ACCURATE
MEASUREMENT**
Because it's hard to hold
the end of the tape on
a flat surface, start the
measurement at the 1-ft.
mark. The extra length also
makes it easier to pull the
tape tightly, which results
in a more accurate
measurement.

Stand the sill plate on edge and trace the outline of the bolt into the plate.

Easy Foundation Bolt Layout

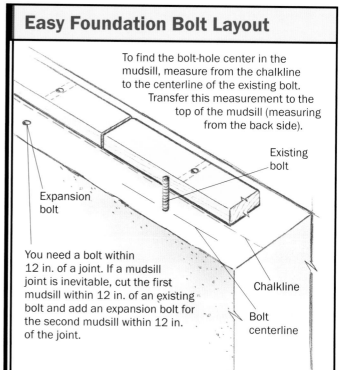

To find the bolt-hole center in the mudsill, measure from the chalkline to the centerline of the existing bolt. Transfer this measurement to the top of the mudsill (measuring from the back side).

Existing bolt

Expansion bolt

Chalkline

You need a bolt within 12 in. of a joint. If a mudsill joint is inevitable, cut the first mudsill within 12 in. of an existing bolt and add an expansion bolt for the second mudsill within 12 in. of the joint.

Bolt centerline

Drill holes in the mudsill as straight as possible. An angled hole will pull the plate off the chalkline. Use a ⅝-in. bit for a ½-in. anchor bolt; place a piece of scrap lumber beneath the plate to protect the bit, or cantilever the mudsill beyond the foundation.

Finding Square with Triangles

According to the Pythagorean theorem ($a^2 + b^2 = c^2$), any triangle with sides that measure 3-4-5 (or any multiple of these) will always have a right angle opposite the hypotenuse (side that measures 5). If a = 3, b = 4, and c = 5, and $3^2 + 4^2 = 5^2$, then 9 + 16 = 25.

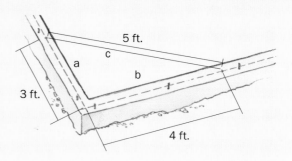

5 ft.

c

a

b

3 ft.

4 ft.

Sources

Metalwest
1229 S. Fulton Ave.
Brighton, CO 80601
800-336-3365
www.metalwest.com

Steel shims
Sokkia
16900 W. 118th Terr.
Olathe, KS 66051
800-476-5542
www.sokkia.com

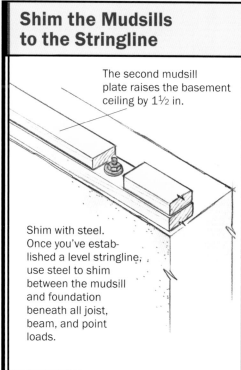

Shim the Mudsills to the Stringline

The second mudsill plate raises the basement ceiling by 1½ in.

Shim with steel. Once you've established a level stringline, use steel to shim between the mudsill and foundation beneath all joist, beam, and point loads.

For a rough count, stack the shims up to the stringline in each location. Steel shims are available in 50-lb. boxes.

Local code requires an anchor bolt within 12 in. of the end of mudsills or of any joints.

Marking the bolt centers on the mudsills for drilling is next (see the right photo on facing page). It's as simple as laying the mudsill alongside the chalkline on top of the foundation, measuring from the chalkline to the center of the bolt, and transferring the measurement to the top of the mudsills (see the top drawing on facing page). At this point, we add insulation (or sill seal if requested) between the foundation and the mudsill.

Shims Level the Mudsill; Bolts Hold It Down

After drilling the bolt holes, two of us place the mudsills over the anchor bolts and another follows behind, adding nuts and wash-

A Builder's Level Finds the High Spots

Commonly though wrongly called a transit, a builder's level rotates only horizontally; a transit rotates both horizontally and vertically. Looking through a builder's level is like looking through a rifle scope, cross hairs and all. Properly set up, the horizontal cross hair represents a level plane, and the magnification is great enough to read a tape measure held 100 ft. away or more. A builder's level is leveled with either three or four thumbscrews and integral bubble vials. Comparing measurements taken in different spots tells you their relative elevations. But this comparison can be counterintuitive. The highest spot, being closest to the level's plane, will have the shortest measurement.

For years, I have used a builder's level to install mudsills, and although I have tried laser levels, I haven't been happy with the results. The Sokkia® E=32 level that I now own cost around $400 in 1999, and it has given me great service.

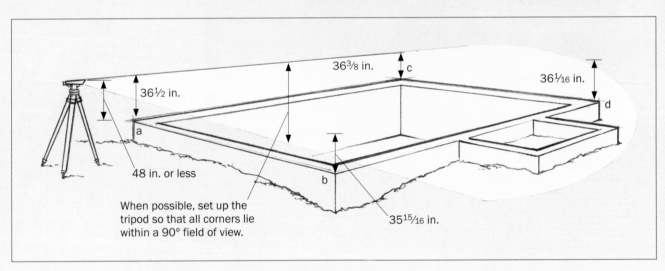

36½ in.

36⅜ in. c

36¹⁄₁₆ in.

d

a

48 in. or less

When possible, set up the tripod so that all corners lie within a 90° field of view.

b

35¹⁵⁄₁₆ in.

ers, tightening them only enough to check for obvious high or low spots. Then we add the necessary expansion bolts at the mudsill joints and nail a second 2x4 on the mudsill. This adds an extra 1½-in. ceiling height in the basement.

Next, we use a builder's level (see the sidebar above) to measure the height of the corners and to look for any high spots. After comparing the measurements, we shim the corners to within ¹⁄₁₆ in. of the highest point on the foundation. Then we run a string

Remove high spots with an air chisel. If a high spot is really bad and it's a short length of wall, an air chisel makes quick work of a labor-intensive job.

Set-Up Tips for Best Results

1. Position the level so that you clearly see each of the foundation's corners within a relatively narrow field of view (90 degrees or less). This helps to eliminate errors associated with swinging the level in wide arcs. Place the level as low to the foundation as possible. Extending the tape or measuring rod high in the air introduces error.
2. With a helper holding a tape, shoot the outside corners a, b, c, and d write their elevations on each corner. The shortest measurement is the high corner (b).

3. Subtract the shortest measurement from each of the other corners, and write the difference (the amount to be shimmed) at each corner.
4. Shim the corners until they measure within $\frac{1}{16}$ in. of the high corner.
5. Run stringlines from one corner to the next. For the areas between the corners, see the photo and drawing on p. 51.

from corner to corner and level the mudsills between.

When shims are necessary, the local building code requires steel shims at joist, beam, and point loads, so I mark these locations on the mudsill. After inserting the shims between the foundation and mudsill,

we tighten the nuts on the anchor bolts and check the height one last time, shooting for plus or minus $\frac{1}{16}$ in.

Jim Anderson is a framing contractor living in Littleton, Colorado.

Framing Walls

■ BY SCOTT MCBRIDE

I've heard a lot in recent years about the speed and the efficiency of California framers, but I find it hard to imagine anyone faster than the Italian-American carpenters who taught me to frame walls in the suburbs north of New York City. These men worked with an extraordinary economy of motion.

I want to discuss wall framing in general and, more specifically, to point out some of the methods and tricks I learned while working with New York carpenters. Even though some of these framing methods differ from those practiced elsewhere in the country, they have worked well for me, and I think they can work for anyone who wants to be more efficient on the job.

Carpentry has a vocabulary all its own. Stud, jack, and header all have meanings outside the carpentry world, but to a framing crew these terms have specific definitions as components of a wall. If you are confused by a sentence that reads, "Toenail the king stud to the bottom plate," then you should familiarize yourself with the drawing on p. 57.

Snapping Chalklines

The first step in any wall-framing method is snapping chalklines on the plywood deck to indicate the locations of the various walls. Wall locations will be shown on the plans. First, I snap lines for all the exterior perimeter walls. If I'm building 2x6 exterior walls— which I usually am these days—I use a 2x6 block to gauge a mark 5½ in. in from the edge of the deck at each end of each wall.

To position the block, I sight down to the corner of the foundation or of the story below, aligning the outside edge of the block with this vertical line of sight (see the photo on p. 56). You can't depend on the rim joist (called the box beam in New York) for registering the block because the rim joist is often warped out of plumb. Staying in line with the true corner is desirable, even if it means a bump in the sheathing at floor level. Otherwise the building tends to grow as it goes up, causing inconsistencies in the span that can complicate the roof framing.

After making a mark on all the corners of the deck at 5½ in., I connect the pencil marks with chalklines. I anchor the end of my chalkline with an awl tapped into the deck. When all the exterior walls are snapped out, I move on to the interior parti-

Nailing the walls together. After all the layout and cutting is finished, the wall components are nailed together. The author nails the doubler to the top plate before the wall is assembled. This then requires that the tops of the studs be toenailed to the plate, instead of through-nailed as in other methods. The pull-out strength of a toenail is greater than that of a through-nail into end grain.

2x6 template. A 2x6 block is used to mark an exterior wall's width on the plywood deck. The author sights down to the floor below to make up for possible irregularities in the alignment of the floor joists.

tions, taking measurements from the plans and transcribing the lines onto the deck. I snap only one line for each wall and scrawl big Xs on the deck with my lumber crayon. The X indicates the side of the line where the wall goes. If there are 2x6 interior partitions as well as ones made of 2x4s, I indicate with my crayon which partitions are which.

Plating the Walls

Plating is the process of cutting to length the bottom and top plates of the walls and temporarily stacking them on the deck (see the photo on the facing page). They can then be marked up to indicate where the various studs and headers will get nailed. In essence, I temporarily put all of the walls in place without the studs in them. My method of framing differs a little from some others in that I cut the doublers now and stack them on top of the other two plates for a three-layer package. Later I'll explain why I do this.

Before cutting any lumber, I think a little about the order in which I want to raise the walls because this sequence determines how the corners of the walls should overlap. Where walls intersect, one wall runs through the intersection. This is called the bywall. The other wall ends at the intersection. This is called the butt wall.

Bywalls have bottom and top plates of the same length that run through the wall intersections. The doubler of a bywall is shorter than the top and bottom plates by the width of the intersecting wall's plates (see the drawing on the facing page). This allows the doubler from the intersecting butt wall to lap the top plate of the bywall. Nailing through the doubler at the lapped corner into the top plate of the intersecting wall holds the walls together.

It is a good idea to cull through your lumber and save your straightest pieces for the longest top plates and doublers. I use the next-best stuff for the bottom plates, which are easier to straighten by nailing to the subfloor. The crooked stuff I cut up for short walls.

There are a couple of things to keep in mind when you are cutting to length the plates and the doublers. Butt joints in the bottom plate can occur almost anywhere. Splices in the top plate—the middle layer—should be offset as much as possible from adjoining walls and beam pockets. Here's why: The integrity of the top-plate assembly —the top plate and the doubler nailed together—depends on having well-staggered joints. An interruption of the doubler is inevitable at wall intersections and beam pockets, so keeping the joints in the middle layer away from these points will maintain good overlap and avoid a weak spot. Splices in the doubler should be kept away from splices in the top plate by at least 4 ft.

If two walls cross each other, you'll have to let one of them run through the intersection and separate the other into two butt walls. The butt-wall doublers can split the overlap, with a joint in the middle of the bywall. Another option is to let one of the butt walls overlap in a full conventional tee. The other butt wall gets no overlap, but instead it is tied to the intersection with a sheet-metal plate on top of the doublers after raising the wall.

To commence plating, the bottom plates are toenailed to the deck on the chalkline, using 8d common nails about every 8 ft.—

Plated and detailed. Information written on the wall plates and the girders defines how the walls are put together. The author also lays out the second-floor joists or roof rafters before assembling the walls. Different-colored ink is used to denote a change in stud length. (Penciled guidelines were drawn on the lumber only for the sake of neatness. It is not common practice for the author.)

The Parts of a Wall

Carpentry has a vocabulary all its own. There is a specific name for every component of a framed wall.

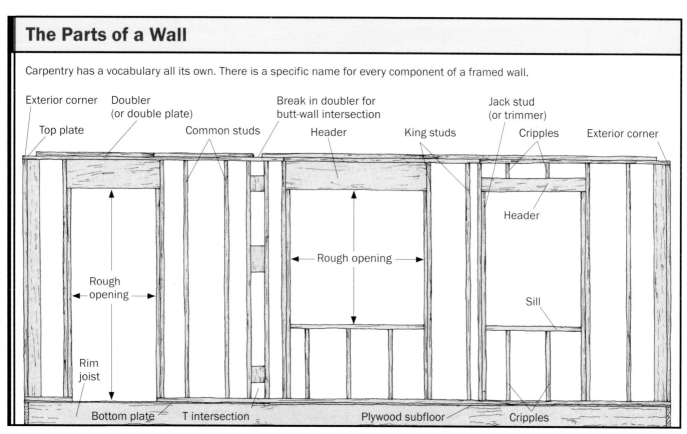

just enough to hold them in position. The top plates—the middle layers—are then temporarily toenailed to the bottom plates.

Finally, I lay down the doublers—the third layer—over the top plates, but instead of tacking them temporarily, I nail them home with 10d common nails staggered 24-in. o.c. I use 10ds here instead of longer nails because the 10ds won't penetrate the bottom plates.

Remember, at the corners where the walls meet, the orientation of the butt joints is reversed, creating the overlap in the doublers that will ultimately lock the walls together. I don't nail the overlap now because I'll have to separate the walls later.

Detailing the Plates and the Doublers

When all the plates are laid down and held together, and the doublers are nailed in place, I'm ready to mark them up, or detail them, with the information my crew and I will need to frame the walls. The first information recorded on the doublers is the width of the door and window rough openings in the exterior walls. The rough-opening marks I make on the top of the doubler are discreet, only about 1½ in. long. I'm saving most of that surface for a later step in the layout.

If windows or doors are shown on the plans dimensioned to centerlines, I measure from the outside corners and mark the centerlines on the outer edge of the doubler. Then I divide the rough-opening dimension of the door or window in half. For example, if the width of the rough opening for a pair of French doors is 6 ft. 4 in., I'll align 3 ft. 2 in. on my tape with the centerline, then mark lines at 0 and 6 ft. 4 in. To check my arithmetic, I turn my tape around end for end. The center should still be at 3 ft. 2 in. I make a V to indicate the rough-opening side of each line (see the photo on p. 57).

Rough openings for interior doors are marked the same on the interior-wall doublers as for exterior doors, but the plans will usually call out the size of the finished door rather than a rough opening. To find the rough opening of a door, I add 2 in. to the width. This allows for a ¾-in. thick jamb and a ¼-in. shim space on both sides.

After locating a rough opening on the doubler, find the length of its header. Openings of less than 6 ft. will require one jack stud, or trimmer, on each side of the opening to support the header. Each jack stud is 1½ in. thick, so the header needs to be 3 in. longer than the width of the opening. Headers over 6 ft. long require double jack studs on each side. That means the length of the header must be 4 times 1½ in., or 6 in., longer than the width of the rough opening.

I use a 2x block as a template to mark the jack locations on the outside of the rough-opening marks. I square the outermost mark down across the stacked edges of the three layers of 2x. This line indicates the end of the header and the inboard face of the king stud. The king stud is the full-length stud to which the jack stud is nailed. On the edges of the top and bottom plates, I show the king stud with an X and the jack stud with an O. For double jacks I use OO. After repeating the process on the other side of the opening, I measure between the outermost marks on the top of the doubler to verify the header length I arrived at earlier by arithmetic. Finally, I write the length of the header on the doubler.

Window headers are marked out the same way as for doors. As far as windowsills and bottom cripples are concerned, I usually come back to them after the walls are up. They aren't needed for structural reasons, and I'm usually in a big hurry to finish the framing and get the roof on. But if I'm going to sheathe the exterior walls before they are tipped into place, I frame below the windows as I go. In that case I'll write the height of the window rough opening on the doubler as well as the width.

I think presheathing pays if you can have the plywood joint even with the bottom of the wall. But some builders, architects, and inspectors require that the plywood joints be offset from the floor elevation to tie the stories together. You can still presheathe in that case by letting the sheets hang over the bottom plate, but I think it becomes more trouble than it's worth. It's usually more economical to let the least skilled members of the crew hang plywood after the walls have been raised.

As soon as I've finished determining all the header sizes, I make a cutlist and give it to a person on the crew who can then get busy making headers while I finish the layout.

Detailing the Doubler

After marking the rough openings on the edges of the top plates and the doubler, the focus shifts back to the top face of the doubler. It's time to lay out the structure that will eventually sit on top of the wall you're about to build. There are very logical reasons to do all this layout now. First, by doing it now, you won't have to spend a lot of time working off a stepladder or walking the plate after the walls are up and the doublers are 8 ft. in the air. The work will already be done. Second, it is easier to align the studs in the wall you're currently building with the loads coming down from above. This is called stacking, and I'll discuss it further later in the chapter.

The structure that sits on top of the doubler may be either a floor or a roof. In either case I start by locating the principle members—girders in the case of floors, ridges in the case of a roof. I mark their bearing positions on the doublers. Then I measure their actual lengths, cut the members, and set them on the doublers. I can now lay out the spacing for the joists or the rafters on the principle beams at the same time I put the corresponding layout on the walls.

To ensure that the principal member is positioned correctly when it's eventually raised, I make a directional notation on the

Layout trick. A flexible wind-up tape is used for layout. Hold a pencil against the tape and swing it across the plate to make marks. The farther the tape is away from the end of the plate, the straighter the lines will be.

beam, such as north end, or driveway side. With the principal members in position, the first joists or rafters I locate are specials, such as stairway trimmers, dormer trimmers, and joists under partitions.

When all the specials have been marked, I lay out the commons. These are individual, full-length framing members (either joists or rafters in this case). The standard spacing is 16 in. o.c., but 12 in. and 24 in. are not uncommon. You want to minimize the cutting of your plywood, so the spacing of commons should result in 8 ft. landing in the middle of a framing member. Assuming 16-in. centers, I hook my tape on the end of the doubler and tick off 15¼, 31¼, 47¼, etc. As an alternative, you can tick off the first mark at 15¼ and then measure the remaining marks from there on exact 16-in. centers.

You can square the tick marks across the doubler with a square or use this trick: Clamp your pencil point against your tape at the given spacing by pinching them together between your thumb and fingers. Swing an arc across the plate (see the photo on p. 59), move your pencil to the next point, repeat, and so on. The error caused by the curvature of the line is negligible, and it diminishes as you move outward. This trick works better with a flexible wind-up tape than with the stiffer, spring-return tape.

By now you're probably wondering if I'm ever going to start nailing the walls together. Be patient, we're almost ready.

Stacking Studs

The spacing for the common studs in the walls is derived from the joist or rafter spacing in the structure that sits above. If joist or rafter spacing is the same as the stud spacing, both 16 in. o.c., for example, I simply extend the layout mark on the top of the doubler down across the edges of the plates. Aligning the framing members like this is called stacking. But if you have, say, joists on 12-in. centers and studs on 16-in. centers, every fourth joist should stack over every third stud. Stacking wherever possible helps to prevent deflection of the top plate, although it could be argued that the presence of sheathing and band joists makes such deflection unlikely. Stacking does facilitate the running of plumbing pipes as well as heating pipes or HVAC ductwork.

On exterior walls, stacking studs from one story to the next makes installation of plywood sheathing more efficient. To support a special joist, such as a double 2x trimmer, I square the layout mark on the doubler down across the edges of the plates and indicate a special double stud.

Over door-header locations, I extend my pencil marks for the cripple-stud spacing across the edge of the doubler and the top plate, but I stop short of the bottom plate. Instead of marking an X here, I mark a C on

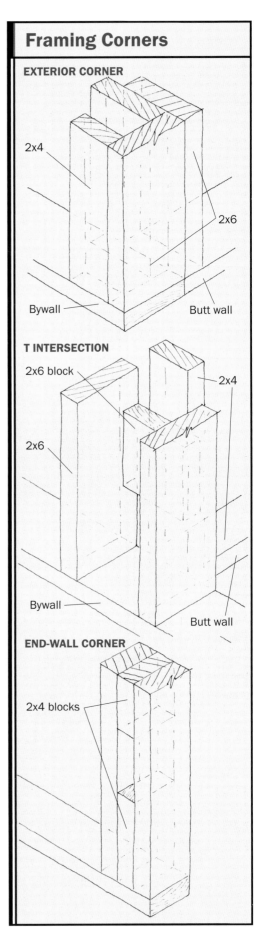

Framing Corners

EXTERIOR CORNER

2x4

2x6

Bywall

Butt wall

T INTERSECTION

2x6 block

2x4

2x6

Bywall

Butt wall

END-WALL CORNER

2x4 blocks

the correct side of the line. This will show the carpenters where to put the cripple studs above the door header. Of course, this step is unnecessary if the header sits tight against the top plate. At window headers I extend the mark across the bottom plate as well, writing C for cripple.

The studs in nonbearing partitions running perpendicular to the joists should also stack, although it's not as crucial as for bearing walls. (A nonbearing partition is a wall that doesn't support a load.) For partitions running parallel to the joists, the placement of the common studs is discretionary. They can be laid out from either end of the wall.

Building Corners

With the stud layout completed, there are only a few more details that need to be mopped up before nailing the walls together. At the end of the exterior bywalls I write CORNER, which means a U-shaped corner unit made up of two 2x6s and one 2x4. This corner design permits easy access for fiberglass insulation (see the drawing on the facing page).

Another step in the final layout is to mark out the studs that go at the end of each butt wall where they intersect a bywall. These inside corners provide nailing for drywall and baseboard and, for that matter, any other type of finish trim that might end in a corner.

Some carpenters preassemble channels to back up these T-shaped intersections. I find it easier to space a pair of studs in the bywall, separated by the flat width of a block. I add the blocks after the walls are raised: one in the middle for 8-ft. walls, two or more for taller walls. By using 2x6 blocks behind 2x4 partitions and 2x8 blocks behind 2x6 partitions, I get a 1-in. space on both sides between the partition and the bywall studs. This provides access for insulation. It also makes nailing drywall and baseboard easier because you don't have to angle the nail as

much to catch the corner stud (see the center drawing on the facing page).

Corner posts for interior 2x4 partitions are made up of intermittent blocks that are sandwiched between full-length studs. This type of corner can also be used at the end of peninsular walls (see the bottom drawing on the facing page). Write B to indicate the blocking.

The final layout step is to number each wall for identification and to indicate the raising sequence. As a convention, I write the number on the left end of the doubler as I look down on it. I then write the same number in front of it on the deck in heavy crayon. These steps help prevent the wall from being installed backward, which is easy to do. I put a slash under the 6 and the 9 to tell them apart. If there are different stud lengths within a story, I write the appropriate length next to the raising-sequence number that has been assigned to each wall. It's not a bad idea to use a different-colored crayon or marker to indicate different nonstandard stud lengths. For example, if I'm writing everything else out in black crayon, I use a red crayon to make the exceptions easy to spot.

Nailing the Walls Together

When the layout detailing is complete, the temporary toenails are removed from the bottom plate only. Each wall is now represented by a separate package containing the bottom plate, the top plate, and the doubler. I stack these packages in an out-of-the-way place on the deck, along with the headers, the corner units, and the principal beams for the structure above. Studs should be leaned against the edge of the deck where they can be reached, but not stacked on the deck. The fastest and most accurate way to mass-produce nonstandard studs is with an improvised double-end cutoff arrangement. Nail two chopsaws or slidesaws to a bench at

Headers and king studs. Tipping the nailed-together doubler and top plate upside down on the deck makes it easy to toenail the header and attach the king stud.

Scribing the jacks. By holding a stud alongside the king stud and against the underside of the header, the jacks can be scribed without measuring. Visible on the bottom plate, half hidden by the carpenter's hand, is a galvanized plate nailed over joints in the plate.

just the right distance apart. Two operators working together lift a stud onto the beds and cut off both ends. This method squares up both ends of the studs and cuts them to length.

When I'm nailing walls together, I usually start with one of the longer exterior walls. I lay the plate package on the deck, parallel to its designated location and pulled back from the edge of the deck by a little more than the length of the studs. You don't want to crowd yourself. I pull out the nails that hold the bottom plate to the top-plate assembly and spread the plates, moving the bottom plate close to the edge of the deck. I'm real careful not to turn the plate end for end.

I find the wall's headers and carry them over to their locations. If the header sits tightly against the top plate, I flip the top-plate assembly upside down and toenail the header down into the underside of the top plate. Then I stand a king stud upside down on the plate and through-nail it to the end of the header with 16d sinkers or 10d commons. I throw a few toenails through the king stud down into the plate as well (see the top photo at left). Now I roll the assembly down flat on the deck.

Some people precut all their jacks, but I like to make them as I nail the walls together. It's simple and fast: I take a common stud and lay it against the king stud, one end butted tightly against the header. Then I strike a line across the bottom end of the king stud onto the jack (see the photo on p. 55) and cut carefully, just removing the pencil line. I nail the jack to the king stud in a staggered pattern, 16 in. o.c. This method of cutting jacks in situ compensates for the variations in header width. The short cutoffs will be used up quickly for blocking.

If the header is offset from the top plate by cripple studs, the wall-framing procedure is substantially the same, except that it's all done flat on the deck, toenailing cripples to the top plate and then to the header (see the photo on p. 57).

Toenailing the Studs

Because the top plate and the doubler are nailed together beforehand, the tops of the studs must be toenailed in place rather than through-nailed. Toenailing requires more skill than through-nailing; it might take a beginner a little longer to learn, but it's not as if it's something he won't have to learn eventually. And toenailing is stronger than through-nailing because it penetrates across the grain. As the walls are jockeyed around on the deck and moved into position to be raised, the bottom plates, which are through-nailed, loosen easily while the toenailed tops hold firm.

If you'd rather not toenail, or if you're using air nailers, which make toenailing difficult, you can tack the doubler to the top plate temporarily for layout purposes. Then pull the doubler off to through-nail the tops of the studs. And finally, nail the doubler back in place using an index mark to ascertain its correct position. Toenailing is the method I was taught years ago, and it's what I'm most comfortable with, so that's the method I'll describe.

When the headers and their jacks and king studs have been nailed in place, I stock the wall with common studs. One end of each stud rests on the top-plate assembly so that it won't bounce around when I start my toenails. I start at one end of the wall, lift up a stud, quickly eyeball it, and lay it back down with the crown pointing to the left (because I'm right-handed). I work my way down the length of the wall until I reach the end (see the photo at right). For 2x6 walls I use 10d commons or 16d sinkers for toenailing. For 2x4 walls I use 8d commons or 10d sinkers. Starting at one end, I start my toenails in the upturned face of each stud—three nails for 2x6, two nails for 2x4. Ideally, the point of the nail should just peek through the bottom of the stud. I work my way down the row.

Bracing the top plate with my feet, I grab the first stud in my left hand. As I shove it away, I turn it 90 degrees to the right so that

Crowning and toenailing. Sighting down each stud determines its crown. All the studs are then laid on the top plates with the crowns facing the same direction. The toenails are then started in all the studs and nailed all at once, production-line style.

it lies on edge, then I pull it back up firmly against the plate. Because the crown now faces up, the stud won't rock on the deck. One blow sets the nail, and two or three more drive it home (see the photo on p. 57). The stud will drift as it's toenailed, depending on the accuracy of the cut, the accuracy of the hammer blow, and the hardness of the wood. Even if you are just a beginning carpenter, you'll quickly learn how far off the mark to start as a way of compensating for the force of your hammer blows.

Through-nailing. As in other methods of framing, the author prescribes through-nailing the bottom of the studs, the jacks, and the cripples. As he works his way down the plate, he aligns each component to its mark and nails it.

Some carpenters drive a toenail in the 1½-in. edge of the stud to hold it in place, but I've always thought this was a superfluous practice that can be dispensed with. When I reach the end of the wall, I double back, firing nails into the other side of the stud—two nails for a 2x6 stud, one nail for 2x4, staggered with those on the other side.

When the studs, the jacks, and the cripples are toenailed to the top plate, it's time to nail the bottom plate. The bottom plate gets through-nailed to the stud with 16d sinkers—three to a 2x6, two to a 2x4 (see the photo above).

Raising the Walls

There are two schools of thought regarding the sequence in which the various walls should be raised. If space on the deck is tight, the long walls must be framed and raised before the other walls have been inflated with studs. Otherwise there won't be enough room. But if I have some room to spare, I'll start with the littlest walls.

I frame the walls and start piling them up. When the pile is finished, the little walls will be on the bottom, and the medium-length walls will be on top. Finally I frame and raise the long exterior walls. Instead of bracing the long walls with a lot of diagonal 2x4s attached to scab blocks, I can pull an adjoining medium-length partition off the pile and drop it into place. This immediately buttresses the long wall. As I work from the perimeter of the house toward the inside, the pile diminishes, and the walls pop up quicker than dandelions.

It may go without saying, but when raising walls, especially long, heavy 2x6 walls, it is important to lift using the power in your legs rather than the smaller muscles in your back. After a wall has been lifted into a vertical position, I ask one person on the crew to line up the bottom plate to the chalkline on the deck while the others hold the wall steady.

The wall can usually be moved to the line by banging on it with a sledgehammer. I nail the bottom plate into the floor framing, one nail per bay. A bay is the space between studs.

All that remains is to rack the walls plumb and brace them. The only braces sticking out into the room should be those necessary to straighten any long, uninterrupted walls that are crooked. The rest of the walls will have been braced by each other. With the next phase of the job already laid out, you're ready to rock and roll.

Scott McBride is a contributing editor to Fine Homebuilding. *He is the author of* Build Like a Pro: Windows and Doors *(The Taunton Press, 2002).*

Framing Gable Ends

■ BY JIM THOMPSON

The first time I had to frame and sheathe the gable end of a house, I did a miserable job. The carpenter I was working for had no system for laying out the studs, which are all of different lengths and have angled cuts on their tops to match the roof pitch. He just said fill in the end between the last rafter and the plate below so that we could put sheathing on it. Then he took off. I did a lot of cutting and fitting and recutting and measuring. The job seemed to take forever.

A few years later another carpenter taught me the method I'm going to explain here. It may look confusing at first, but once you get it, you'll have a gable end framed in less time than it takes to read this article.

The studs are lined up over the wall studs below (called stacking) and notched around the gable-end rafter (see the drawings on p. 66). The important part of the notch is the pitch cut, which is on the same angle as the rafter (see the right drawing on p. 66). When the studs are cut and installed this way, they flush up to both the interior and exterior of the framing. I like the solid feel of gable walls that are framed with this method.

Before installing the gable-end studs, a little preparation is required. You should place a couple of sheets of plywood on the ceiling joists for a work area. On this work area you need a circular saw, a level that's at least 4 ft. long, a framing square, and enough studs and sheathing to cover the gable. It's also helpful to leave out the next-to-last rafters to give yourself adequate room to work. I only forgot that trick once.

Story Pole

Begin the job by selecting a straight stud that is long enough to extend above the top of the rafter at your highest stud. This is your story pole, and it eventually will have the height of the notches for all the gable studs marked on it. Draw a big R on one edge to indicate the right side of the gable, and a big L on the opposite edge to record the left side. Then put an arrow on one edge to show which way is up.

Begin marking at the high end of the gable by putting the bottom of the story pole directly over the stud nearest the peak. As shown in the drawing on p. 66, the story pole is held against the rafter at the top and plumbed with the level. Next make a mark on the rafter to show the uphill side of the

Mark the Layout with a Story Pole and Level

NOTCHED GABLE-END STUD

STACKED STUDS

If a house is to be sheathed with plywood or OSB, its gable-end studs need to be aligned with the studs of the walls below. A quick way to mark the layout of the studs on the rafters is with a story pole, and a level at least 4 ft. long, as shown below.

Mark side of rafter.

Pitch cut

Mark edge of story pole.

Plumb story pole with level.

Left side

Gable-end rafter bears on notched studs.

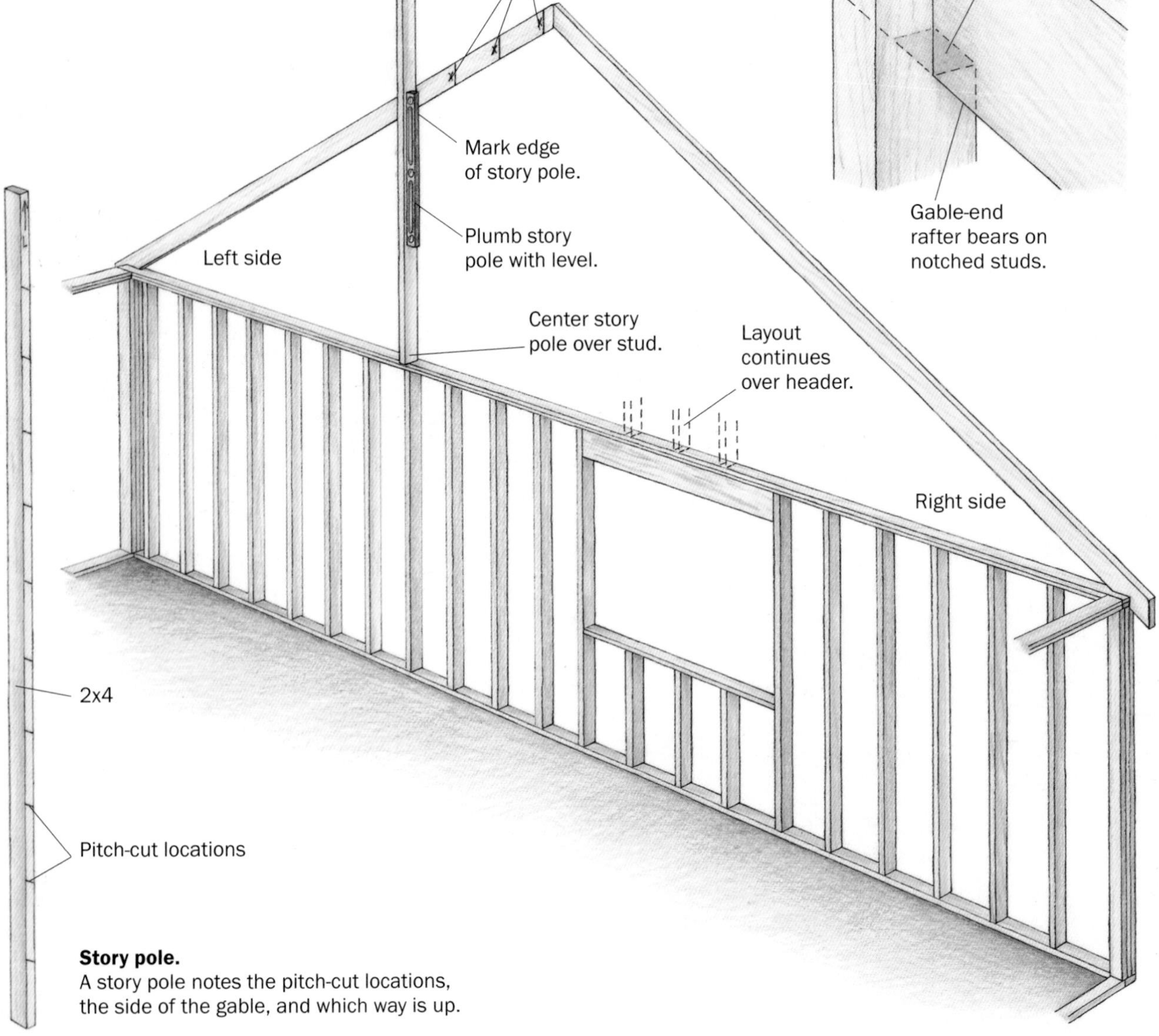

Center story pole over stud.

Layout continues over header.

Right side

2x4

Pitch-cut locations

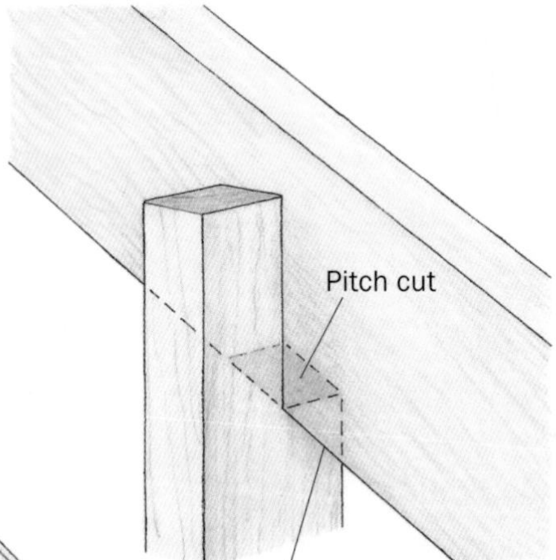

Story pole.
A story pole notes the pitch-cut locations, the side of the gable, and which way is up.

story pole and another mark on the edge of the story pole that records the bottom of the rafter. With the marks made, pull the story pole away from the rafter and draw an X on the downhill side of the line on the rafter to show where the stud will go.

Repeat this procedure for all of the studs on one side of the gable, then turn the story pole around and do the same for the studs along the other side. When you're finished, your rafters should be marked for stud layout, and your story pole should have similar markings on both sides that show the positions of the pitch cuts (see the left drawing on this page).

Making the Cuts

Now spread out enough studs to do the right side of the gable. Lay the studs on edge, snug them up to the story pole and use a framing square and a sharp pencil to transfer the story-pole marks to the studs (left drawing, this page).

The marks on the edges of the studs are registration points for your first cut. Arrange the studs so that the marks line up (middle drawing, this page) and run the circular saw down the line with the depth set equal to the thickness of the rafter (typically 1½ in.).

The next step is to cut off the tops of the studs so that they don't extend above the rafters. I leave 3½ in. above the pitch cut. That's enough material to anchor a nail—any more wastes wood. The wide side of my saw's base (Skil® 77) is 3½ in. from the edge to the blade. So I eliminate measuring and use the saw's base to tell me how far from the pitch cut to make my crosscut. As I make my crosscuts, I set aside the offcuts. They will be the studs for the left side of the gable.

Now I can cut out the waste to make the notches. Because the narrow side of my saw's base is 1½ in. wide, I use the edge of the base to line up my cut (bottom right drawing, this page). Resist the temptation to hold the stud in one hand while operating the saw with the other hand. Instead, tack or

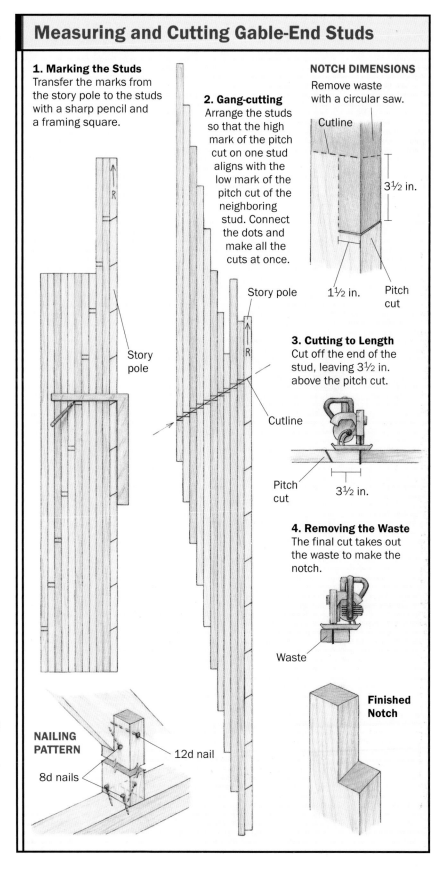

Measuring and Cutting Gable-End Studs

1. Marking the Studs
Transfer the marks from the story pole to the studs with a sharp pencil and a framing square.

2. Gang-cutting
Arrange the studs so that the high mark of the pitch cut on one stud aligns with the low mark of the pitch cut of the neighboring stud. Connect the dots and make all the cuts at once.

Story pole

Story pole

Cutline

NOTCH DIMENSIONS
Remove waste with a circular saw.

Cutline

3½ in.

1½ in.

Pitch cut

3. Cutting to Length
Cut off the end of the stud, leaving 3½ in. above the pitch cut.

Cutline

Pitch cut

3½ in.

4. Removing the Waste
The final cut takes out the waste to make the notch.

Waste

Finished Notch

Pitch cut

NAILING PATTERN

12d nail

8d nails

Cutting Gable-End Studs Using the Common Difference

Most roofs I have worked on over my career as a rafter cutter have been gable roofs. And most of the houses I've worked on here in southern California have been finished with stucco—not plywood covered with siding as in other parts of the country. Instead of relying on plywood or OSB for diagonal braces, builders here use 1x6s that are let into the studs to brace the walls. As a consequence, the studs on a gable end don't have to line up with the studs in the wall below to continue the layout for sheathing. The gable-end studs simply have to end up on 16-in. centers to provide ample backing for the stucco.

The method I'll describe here is used by a production framer to determine the lengths and the pitch cuts of the gable-end studs by means of the common difference, which is the consistent incremental change in length of equally spaced studs under a rafter. With this method, I make a minimum number of cuts, and I can work on the ground next to the lumber pile.

Roof Pitch

Designers typically show the pitch of a roof by drawing a right triangle on the elevation or the roof plan. The base of the triangle represents 12 horizontal inches (the run), and the vertical leg of the triangle shows the amount the roof rises in that horizontal foot. The hypotenuse of the triangle therefore represents the pitch of the roof. This example has a 6-in-12 pitch, and the run of the roof is 15 ft.

To calculate the lengths of the gable studs, first find the length of the longest stud. Here's the equation. Multiply the pitch (6 in.) times the run (15). The answer is 90 in. (see the top drawing on p. 70). A 90-in. stud would be necessary to support the rafter at its highest point. But there's a curveball. A gable end typically has a vent in it, so the studs on both sides of it must be moved over a bit so that they don't interfere with the vent. To allow for the vent, I deduct half the pitch from the longest stud. That reduces the stud by 3 in. to 87 in. long, and moves the stud over 6 in.

What's the Difference?

The centerlines of gable studs are normally spaced 16 in., which is equal to $1\frac{1}{3}$ ft. To find the difference in length between adjacent studs, multiply $1\frac{1}{2}$ by the pitch. In my example the equation is $1\frac{1}{3} \times 6 = 8$ in. difference between studs.

This gable end needs eleven 8-ft. 2x4s. Two studs are going to come out of each one of these 2x4s—one for the right side of the gable and one for the left side. The studs will be mirror images of each other.

On a flat work surface, place the 2x4s on edge with their ends square to one another. Measure 87 in., and mark the 2x4 closet to you at that clamp the stud to a work surface to hold the stud steady during the cut.

When you flip over the story pole and line up the offcuts for marking the left side of the gable, you'll notice that most of the offcuts are usable. In this example, the job requires only two additional studs, and the story pole can be used for one of them. This economy of material will vary with the pitch of the roof, the layout of the wall below, and the length of the wall, but I've found the method to generate very little waste.

Stud and Sheathing Installation

To install a stud, set its bottom on the plate and align it with the stud below. (I prefer to start with the tallest stud and work toward the eaves, but it doesn't really make any difference.) Now slide the notch under the rafter until it is just snug. Be careful not to push too hard because you can easily put a point with a pencil. Then mark each successive 2x4 8 in. shorter.

Cutting the Rafters

A gable-end stud tucks under the rafter, so its end has to be cut at the same pitch as the rafter. There are two quick ways to find the correct angle. With a framing square, align the edge of a 2x4 with the 6-in. mark on the tongue of the square and the 12-in. mark on the blade. Mark the angle with a pencil. Set the table of your circular saw on the face of the 2x4 and adjust the angle of the table so that the blade matches the pencil line. Or your can look in a book on rafter cutting to see what angle a 6-in-12 pitch equals. The answer is 26½ degrees. The book I use is *Full Length Roof Framer* (A.F.J. Reichers, P.O. Box 405, Palo Alto, CA 94302).

When you cut a stud with the saw set at an angle, the resulting mitered end gives the stud a long point and a short point. I make my measurements to the long points. And when I cut any material on a miter, I cut from the side that will allow my mark to be the long point of the cut. To finish marking the gable studs, measure the length of the shortest stud. In my example, it's 7 in. long. As shown in the drawing, add that length to the 2x4 that has the longest stud laid out on it. The long point of the shortest stud will correspond to the short point of the longest stud.

With a framing square aligned on the mark that defines the bottom of the shortest stud, mark the waste cut across the rack of 2x4s with a pencil. Make this cut with the saw set for a square cut. Now set the saw table at 26½ degrees and cut each 2x4 at the cutline marks. When you're finished, hold up the two shortest studs. They should be the same length.

Nailing

I start with the longest studs, and I use a level to make sure they're plumb. The studs should just touch the bottom of the rafter—don't wedge them in place, or you'll put a hump in the rafter.

After putting in the longest stud on one side of a gable, take one that will fit about in the middle of the rafter. Look down the rafter to see if it has a hump or a sag in it. Move the stud up or down to take out any hump or sag. Don't drive the nails through the stud into the rafter. Instead, toenail a couple of 16d nails throughout the outside face of the rafter into the tops of the studs.

Once you've got the middle stud in, fill in the remaining ones. Use the level to keep the long studs plumb. It's pretty easy to plumb the short ones by eye.

Elmer Griggs is a former carpentry instructor for Ventura County and Los Angeles County apprenticeship committees. He lives in Reseda, California.

hump in the rafter with the first stud. Then none of the studs will fit correctly.

Toenail the bottom of the stud to the top plate with two 8d nails on one side and a single 8d nail on the other side. Nail the top with a 12d through the top of the stud into the rafter. Then add an 8d toenail into the bottom of the rafter from the uphill side of the stud.

I prefer to sheathe a gable end from inside the building, so I rip my sheathing into 2 ft. by 8 ft. pieces. That's because it's easy to lean out and nail off the bottom of a 2-ft. wide piece of sheathing, but it's tough to reach the bottom of a 4-ft. wide piece. (In areas subject to earthquakes or high winds, this method may not fulfill shear-wall requirements.)

I start the first piece of sheathing so that its edge lands on the center of a stud that's on my 16-in. o.c. layout. Not all studs are placed at layout; for example, a gable-end vent may require studs to fall outside the layout.

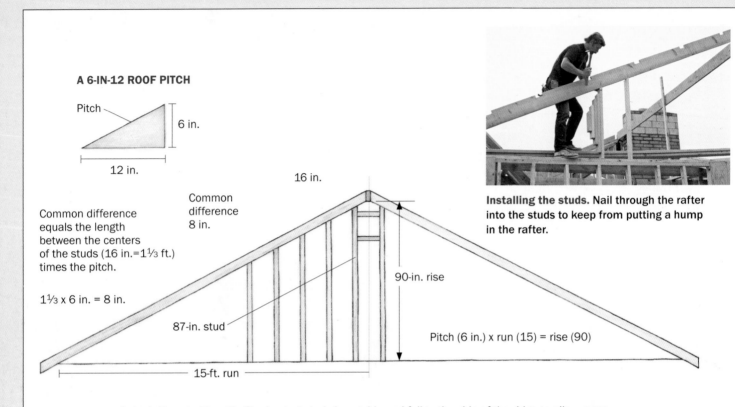

A 6-IN-12 ROOF PITCH

Pitch — 6 in.

12 in.

Common difference equals the length between the centers of the studs (16 in.=1⅓ ft.) times the pitch.

1⅓ x 6 in. = 8 in.

Common difference 8 in.

16 in.

90-in. rise

87-in. stud

15-ft. run

Pitch (6 in.) x run (15) = rise (90)

Installing the studs. Nail through the rafter into the studs to keep from putting a hump in the rafter.

Calculating stud length. The longest studs in a gable end fall to the side of the ridge to allow room for the gable vent. The author calculates the length of the longest studs by multiplying the pitch of the roof by the run of the rafter minus half the pitch.

Pitch (6 in.) x run (15) – ½ pitch (3 in.) = longest stud (87 in.)
6 in. x 15 - 3 in. = 87 in.

Save the cutoffs from the pieces of sheathing that project above the rafters. You can use them for starting or finishing a course because the cutoffs already include the pitch cut. When your sheathing gets too high to reach over, nail a 2x4 across the gable-end studs to make a place to stand. Use two 16d nails in each stud for this piece. When you're finished, leave the 2x in place because it helps stiffen the gable end.

But What About . . .

Here are answers to the typical questions I get about this method. First, if your centers are regular, and the slope of the roof is constant, why not just mark the first stud, get the common difference and mark the studs that way? Some folks prefer that technique (see the sidebar above), but I have found that it really takes no time at all to mark the story pole, which allows me to lay out the

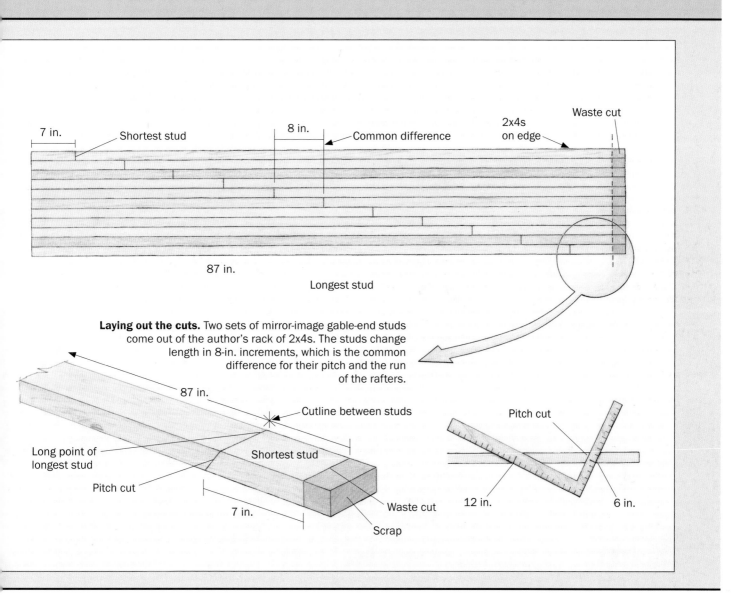

7 in.

Shortest stud

8 in.

Common difference

2x4s on edge

Waste cut

87 in.

Longest stud

Laying out the cuts. Two sets of mirror-image gable-end studs come out of the author's rack of 2x4s. The studs change length in 8-in. increments, which is the common difference for their pitch and the run of the rafters.

87 in.

Cutline between studs

Pitch cut

Long point of longest stud

Pitch cut

Shortest stud

7 in.

Waste cut

Scrap

12 in.

6 in.

stud position on the rafter at the same time. Having the rafter laid out helps avoid bowing it upward if the cut is a little off, and the stud is too tall or too short.

Second, why not lay out the right and left halves from the same marks? Unless your layout below is exactly centered on the peak, which I've never found to be the case, it won't work.

Probably the most important things to remember while using this method are the

orientation of the story pole while transferring the marks and which direction to slide the studs when you're preparing to make the pitch cuts. You simply have to remember which end is up.

Former framing carpenter Jim Thompson is now a structural engineer with McCormac Engineering Associates in Ellicott City, Maryland.

Straightening Framed Walls

■ BY DEREK MCDONALD

As a craftsman, I'm offended when I notice bowed walls and wavy ceilings in newly finished houses. Everyone knows it's tough to find straight, knot-free lumber these days. But bad lumber is not the only cause of bad walls. Extreme weather conditions that strike before the roof is dried in, as well as fluctuations in temperature and humidity afterward, can make even good studs go bad.

The frequency of warped and twisted studs in the average frame house can be reduced if lumber is kept banded and covered until the framers are ready for it. Conscientious framers will also crown moderately bowed studs and cull the worst offenders, setting them aside for use as nailers and blocking. Unfortunately, the supply of skilled labor seems to be dwindling faster than that of straight lumber; so my company maintains a crew of "pickup" carpenters like me who follow behind the framers, straightening studs and flattening walls.

Tolerances Vary from Room to Room

I work for a high-volume framing contractor who builds tract houses in California, so I have to balance my perfectionist tendencies with the pressure to get the job done as efficiently as possible. Keeping this balance requires me to choose which walls are most critical and will thus receive more of my attention. The choice is not difficult; entries and long hallways are more visible than bedrooms, closets, and garages, and are therefore held to a higher standard. More critical are bathrooms—where cabinets and mirrors must lie flat—and kitchens. Because their long rows of cabinets and countertops make flat walls and straight corners essential, kitchens are the most critical rooms of all. For kitchen walls, I allow no more than $\frac{1}{16}$-in. variance from perfectly straight and flat, but I'll accept as much as $\frac{3}{16}$ in. for the garage and the closets.

Efficient Straightening Requires a Good Eye

To minimize stud deflection, I prefer not to begin straightening walls until all roofing tile is in place and until bearing walls are fully loaded. Once I start working, all my measuring is done by eye. My tape measure never leaves the toolbox, and I rarely use a chalk box or dry line.

Trust your eye. A visual check gives a good indication of straightness. In most cases, when all of the studs fall crisply into line, the wall is acceptable.

I use the top and bottom plates as starting points to check the straightness of a vertical plane, so my first step is to verify that both ends of each stud are flush with the plates. Any stud that isn't where it's supposed to be gets hammered into position and toenailed. With the starting points in alignment, I give myself a glimpse of the work ahead by sighting down the length of the wall (see the photo above) and visually lining up the studs. A quick glance such as this one lets me know what gross irregularities are lurking, and when all the studs line up like soldiers, I know there's a good chance I can move on to the next wall.

Unless every stud lines up perfectly, the next step is to find out where the deviants are. This step requires a long straightedge. Any perfectly straight piece of wood or steel will do, but because I believe a lightly armed

On those rare occasions when an offending stud cannot easily be planed, shimmed, or simply removed, more elaborate solutions are called for. "Strong-backing" (photo above) is one technique to force a severely bowed (more than 3/8 in.) stud back into alignment and keep it there. A bowed corner stud (photo right) must be temporarily cut loose and then wedged out from the corner before it can be planed straight.

carpenter is an efficient one, I use the same 8-ft. box level that I use for plumbing walls. I've also found that the smooth surface of my box level is much kinder to my fingers after a day of continuous handling than the rough edges of wood or the sharp edges of steel.

Repairs Are Noted on Stud Face

I prefer to straighten one wall at a time: eyeballing, making note of problems, and performing necessary surgery before moving on to the next wall. Starting in a corner and working my way out, I place the straight-edge vertically against each stud and trust my eye to judge the amount of correction needed, and where (see the photo below). Using a simple shorthand we've devised, I mark the remedy directly on the face of each

A little daylight is a bad, bad thing. Using a straightedge that runs from top plate to bottom plate, the author can easily determine how much a stud deviates from the ideal. Because it's part of a highly visible hallway, this stud will need a shim.

A prescription for change. The author checks every stud with a straightedge and marks the remedies on the stud faces as he goes. An S-shaped squiggle indicates that the stud needs to be planed. Hash marks indicate the number of shims needed.

offending stud using a lumber crayon. If the stud bows outward, I squiggle an S-shape over the area that needs to be planed. If the stud bows inward, I make one or more hash marks in the center of the bow to indicate the number of cardboard furring strips I estimate I'll need to fill in the gap (see the photo at left above).

If I find a stud that needs more than 3/8 in. of planing or furring, I put a big X on the face. This note reminds me to come back later with a reciprocating saw, cut the nails that anchor the stud to the plates, and replace it. Sometimes it's difficult to replace one of these bad studs—it could be part of a complicated framing scheme, or maybe the electrician has beaten me to the job and run wires through the walls. In this case, we straighten the stud using a technique we call strong-backing.

Strong-backing is accomplished by notching the offending stud at midpoint (on the face that bows outward) to let in a 2x4 or 2x6 on the flat. We cut a block to fit between the two studs that flank the bad one; then we nail one end to one of the flanking studs and use the block as a lever to draw

Planing begins at the midpoint. To straighten a bowed stud without measuring, the author starts with a foot-long pass in the middle and overlaps with increasingly longer passes.

the offender back into the plane of the wall. When the offender has been reformed, we nail the free end to the side of the other flanking stud (see the top photo on the facing page).

Plane First, Ask Questions Later

After each stud has been checked with a straightedge, I use my planer to take off high spots. Unless I know that a knot high or low on the stud is causing the bow, I begin planing at the midpoint of the stud. Starting with a foot-long pass, I overlap with increasingly longer passes until I have planed most of the stud's length (see the photo at right above). The severity of the bow determines the number of passes. When I think I've removed enough material, I get the straightedge and check my work.

Planing is a straightforward procedure unless the offender is a corner stud. An outwardly bowed corner stud must either be replaced or temporarily wedged far enough from the adjoining wall's corner stud to allow room for the planer (see the bottom left photo on p. 74). If planing is the only solution, I first drive a small wooden wedge between the afflicted stud and the stud to which it is nailed. This step allows room to slide my reciprocating saw's blade between the two studs to cut the nails holding them together. Then I force the corner stud out by driving two thick wedges between the studs, one about 2 ft. from each end. After planing the bow, I remove the wedges and renail the stud.

How do you spell relief? Available in bundles of 50, precut cardboard shims make quick work of a tedious job.

An efficient shimming operation. Duct tape and a scrap piece of drain pipe create a handy quiver to ensure that the cardboard shim stock is always within the author's reach.

Precut Shims Make Life Worth Living Again

After I've planed the high spots, I shim the low spots. Shimming studs used to be a tedious process of trial and error, but that changed a few years ago when we started buying precut cardboard shim stock (see the photo at left above). I'm told there are different varieties of precut shims available now; the ones we've used are 45-in.-long, ¹⁄₁₆-in.-thick cardboard strips that come bundled in groups of 50.

The procedure I use for shimming is similar to the one I use for planing. Often, a single shim is sufficient. But if the bow is a pronounced one, I start by laying down a short (1 ft. or 2 ft. long) strip over the midpoint of the bow. Then I work outward from the center of the bow, overlapping (by various amounts) successively longer strips to achieve a blending effect. I fasten the shim stock to the studs using a hammer tacker loaded with ³⁄₈-in. staples (see the photo at left).

After I've planed and shimmed all the studs in the wall, I need to check the plane of the entire wall. At this time, I place the straightedge horizontally across the studs, at various points up and down the wall, to see how the framing surfaces flow together (see the photo on the facing page). If adjacent studs vary from plane more than the

fraction of an inch I allowed when I was straightening individual studs, that means I still have to do a bit of planing or furring to fine-tune the wall.

Don't Plane a Truss without Prior Approval

Most of my effort goes toward straightening walls, but I don't ignore ceilings. I follow essentially the same procedure for ceilings as I do for the least critical walls. Large rooms can be a problem, though, because an 8-ft. straightedge is too short to give a true idea of straightness. So when my eye picks out a serious irregularity, I pull a dry line from one end of the suspect framing member to the other. An upward bow is easily shimmed to meet the line. A downward bow is another story.

If the framing members were solid-sawn joists, I could snap a chalkline on the side of the joist and plane to the line. Unfortunately, I work with trusses. Although the engineer who designs our trusses will allow me to plane up to ¼ in. off the bottom chord, he prefers that I "paper down when feasible." What he means is that rather than plane up to the line, I should shim down to the low point. Because this procedure lowers the surface of the truss in question beneath that of its neighbors, I then have to draw my straightedge across the chords and shim the adjacent framing surfaces enough to feather out the differences.

Use Your Illusion

Whether the subject is walls or ceilings, it's important to keep in mind that I'm hardly ever trying to create perfectly flat surfaces. In most cases, my goal is to create the illusion of perfection; I do it by straightening and aligning, as much as the laws of physics and time allow, and by blending and creating smooth transitions where they do not exist. Consider, for example, a long wall

Checking the plane of the wall. After the offending studs have been planed and shimmed for vertical straightness, the author then checks across the studs to see how they line up horizontally.

in which 11 of 14 studs have a slight but consistent inward bow. By planing the three straight studs to imitate the shape of the others, I can reach a compromise between aesthetics and economics. The other option—furring all 11 bowed studs—would take twice as much time, and no one would know it but me.

When he's not surfing the pipeline, **Derek McDonald** *works as a carpenter for HnR Framing Systems Inc. in Poway, California.*

Sources

Fortifiber™ Co.
55 Starkey Ave.
Attleboro, MA 02703
800-343-3972
www.fortifiber.com
pre-cut shims

Framing a Roof Valley

■ BY RICK ARNOLD

Framing a complex roof is one of the trickiest parts of home building. But it doesn't have to be. If you begin with accurate, as-built measurements and use a construction calculator to do the math, you can cut all of the rafters for most roofs on the ground in just one shot.

This roof valley is a perfect example. To understand the concept and how all the pieces are laid out and cut, I picture the roof two-dimensionally, in plan view (see the drawing on p. 80). Then I use a construction calculator to find the correct length of each rafter.

To get started, I need to know only two things. The first is the pitch of the roof. On this project, the roof is a 12 pitch, or 12-in-12. The second thing that I need to know is the total run of the common rafters (see the drawing on p. 80).

As a house is framed, measurements can vary slightly from the design, so I disregard the plans when finding the run of the common rafters, and instead take my own measurements. I measure the width of the building first, including the sheathing. The run of the common rafters is equal to half of this measurement, less half the thickness of the ridge board. Once I have the run of the common rafters, I use the calculator to find the lengths of all the rafters (see the examples on pp. 82–83).

The Valley Rafter Has a Different Pitch

The run of the valley rafter is longer than the run of the common rafters, so it takes 17 in. of run for the valley rafter to rise the same distance that the common rafters rise in 12 in. Therefore, the pitch of the valley rafter is 12-in-17.

The first cut that I make on both the common rafters and the valley rafter is the plumb cut, where the rafter will meet the ridge board. On the common rafters, the plumb cut is marked with a framing square for a 12-in-12 pitch (45 degrees) and cut square (i.e., no bevel) (see the drawing on p. 83). The valley rafter is marked for a 12-in-17 pitch and then cut with a double bevel to fit into the ridge intersection (see the top photos on p. 83). Because most framing lumber is not perfectly straight, I check each piece and mark the crown before I start cutting the lumber into rafters. Keeping all of the crowns pointing up prevents waves from marring the finished roof.

The Valley Rafter Is the Diagonal Where Intersecting Roofs Meet

With a regular valley, where the walls meet at right angles and the roof pitches are the same, the valley rafter intersects the building diagonally at 45 degrees; hence, the jack rafters intersecting the valley require a 45-degree bevel cut. The plan view makes it clear that the run of the valley rafter is longer than the run of the common rafters.

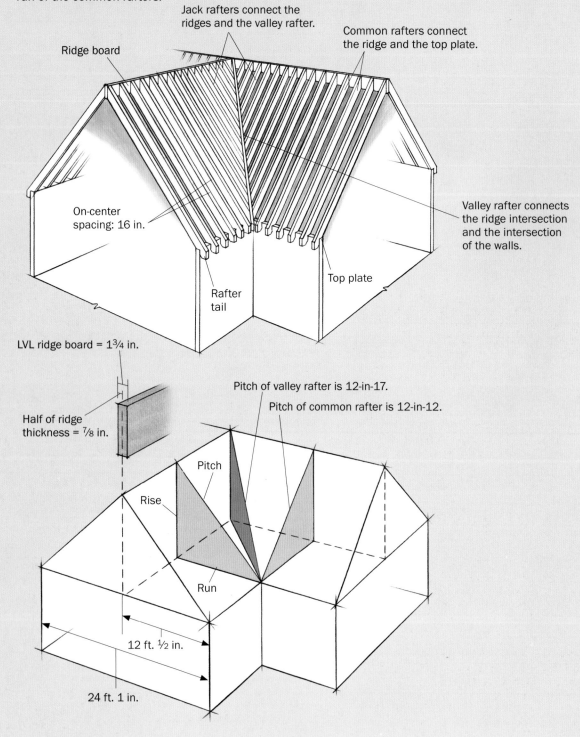

Jack rafters connect the ridges and the valley rafter.

Common rafters connect the ridge and the top plate.

Ridge board

On-center spacing: 16 in.

Valley rafter connects the ridge intersection and the intersection of the walls.

Rafter tail

Top plate

LVL ridge board = 1¾ in.

Half of ridge thickness = ⅞ in.

Pitch of valley rafter is 12-in-17.

Pitch of common rafter is 12-in-12.

Pitch

Rise

Run

12 ft. ½ in.

24 ft. 1 in.

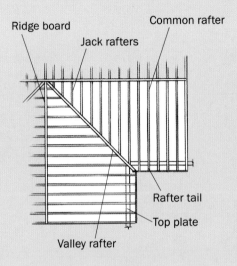

Ridge board

Common rafter

Jack rafters

Rafter tail

Top plate

Valley rafter

1. DETERMINE THE RISE, RUN, AND PITCH
Because the run of the valley rafter is longer than the run of the common rafters, but the rise is the same, the pitch of the valley rafter changes from 12-in-12 to 12-in-17 in this case.

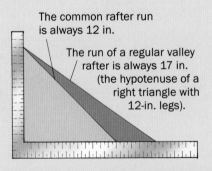

The common rafter run is always 12 in.

The run of a regular valley rafter is always 17 in. (the hypotenuse of a right triangle with 12-in. legs).

The run of the common rafters is equal to half the width of the house. But because the rafters are nailed to the face of the ridge board, and the ridge board bisects the house, half the thickness of the ridge is subtracted from the run.

Find the Run of the Common Rafters
12 ft. 1/2 in. (half building width) – 7/8 in. (half of ridge thickness) = 11 ft. 11 5/8 in. (run of common rafters)

The Valley's Bird's Mouth and Tail Are a Little Tricky

The key to cutting the tail end of the valley rafter is to make sure it will work with the common rafters to form the plane of the roof, soffit, and fascia (see the bottom drawing on p. 83). Therefore, the layout of the common rafters acts as a starting point. On both the common and valley rafters I measure the length along the top of the rafter and mark a plumb line. To form the bird's mouth—a triangular shape cut in the bottom of the valley rafter where it sits on the top plate—I mark a line for the seat cut, perpendicular to the heel cut. On the common rafters, the seat cut is the same length as the top plate and sheathing. On the valley rafter, the seat cut is located to maintain the same height above plate (HAP) as the common rafters (see the drawing on p. 83).

In a perfect world, the heel cut on the valley rafter would mimic the ridge cut and form a point that fits into the intersection of the top plates. However, with a single-member valley rafter, I don't bother with this cut. (Valley rafters are structural members of the roof, so sometimes they are doubled up. Consult an engineer if you are unsure how to size a valley rafter.) Instead, I simply extend the heel cut so that its sides fit snugly against the sheathing. To form a consistent plane for the soffit and fascia, the length of the overhang on the valley rafter is adjusted so that the tail projects the same distance from the house as the tail of the common rafters. To align the tail of the valley rafter with the tail of the common rafters, I simply extend the overhang half the thickness of the valley rafter and then cut the tail with a 45-degree bevel (see the drawing on p. 83).

Rafter Layout

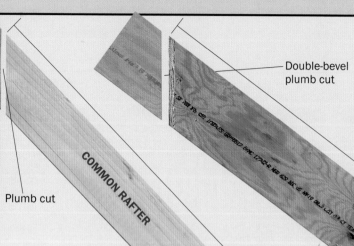

Double-bevel plumb cut

Plumb cut

COMMON RAFTER

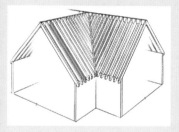

Find the Length of the Common Rafter
Enter: 11 Feet 11 Inch ⁵⁄₈
Press: Run
Enter: 12 Inch
Press: Pitch
Press: Diag
Result: 16 ft. 11⅛ in. (length of common rafter)

2. LAY OUT THE COMMON RAFTERS

Make a plumb cut that reflects the 12-in-12 pitch of the roof on one end of the rafter. Then measure the length (see calculation at right) from the long point of the plumb cut and mark a plumb line to locate the heel cut of the bird's mouth. The seat cut is perpendicular to the heel cut and equal to the width of the top plate and sheathing. The overhang, taken from the blueprints, is measured perpendicular to the heel cut.

A Quick Way to Make the Double-Bevel Plumb Cut
Mark two plumb lines the same distance apart as the thickness of the valley rafter. Set the saw to 45 degrees and make the first cut so that the outside line becomes the long point. Cut the inside line in the other direction but with the same bevel.

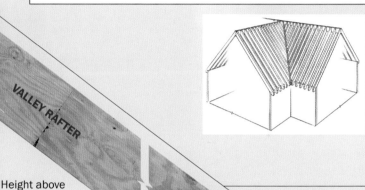

3. LAY OUT THE VALLEY RAFTER

The plumb cut on the valley rafter, also called the ridge cut, is marked for the 12-in-17 pitch of the valley rafter and cut with a double bevel to fit into the intersection of two ridge boards (see above). Starting with the length of the common rafters, find the length of the valley rafter on the construction calculator (see below).

Find the Length of the Valley Rafter
Enter: 11 Feet 11 Inch $^5/_8$
Press: Run
Enter: 12 Inch
Press: Pitch
Press: Diag
Result: 16 ft. 11$^1/_8$ in. (length of common rafter)
Press: Hip/V
Result: 20 ft. 8$^3/_4$ in. (length of valley rafter)

Height above plate

The seat cut and the heel cut are known collectively as the bird's mouth.

Height above plate from common rafter

········· Top plate ·········

Heel cut

Seat cut

Overhang

Beveled fascia cut

Plumb cut for fascia

Heel cut moved $^7/_8$ in. to clear wall intersection

Level cut for soffit

Soffit cut

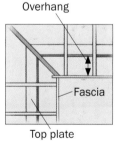

Bevel the Valley Rafter's Tail
Cutting a 45-degree bevel on the tail of the valley rafter creates a place to nail the intersecting fascia boards.

Overhang

Fascia

Top plate

4. LAY OUT BIRD'S MOUTH AND RAFTER TAIL

The valley rafter is measured and marked with a plumb line for the heel cut. The seat cut is located along the heel-cut line so that the height above plate is consistent with the common rafter. The heel cut is moved half the thickness of the valley rafter ($^7/_8$ in.) so that it fits into the intersection of the top plates. The end of the rafter gets a beveled plumb cut for the fascia and a level cut for the soffit. The overhang of the valley rafter is calculated as the diagonal of the common-rafter overhang.

Find the Length of the Valley Overhang
Enter: 10 Inch $^1/_2$ (length of the common overhang)
Press: Run
Enter: 45
Press: Pitch (diagonal angle of the valley rafter, not the pitch of the roof)
Press: Diag
Result: 14$^7/_8$ in. (overhang measurement for valley rafter)

Jack Rafters Are Beveled to Meet the Valley Rafter

The tops of the jack rafters are marked and cut identical to the common rafters, but the bottoms have a 45-degree beveled, or cheek, cut that fits against the diagonal valley rafter. For each length of the jack rafters, I cut two rafters with opposing bevels on the cheek cut (see the drawing at right). I use the first jack rafter as a pattern to mark another one the same length (see the photo at right). But when I cut the cheek on the second rafter, I cut in the opposite direction, reversing the bevel.

Keep the Valley Rafter Straight During Installation

With the ridges in place, the valley rafter is nailed into position. It is important to double-check the HAP with the valley rafter in position. If the HAP is short, it may be possible to adjust it with a shim between the top plate and the seat cut. If it is long, the seat cut will have to be adjusted. The valley rafter then is nailed into the ridge intersection and into the top plate. To straighten out the valley rafter and keep it straight, I first install a pair of jack rafters about halfway down the valley (see the photo on p. 86). If there is more than one valley, I install jack rafters at the midpoint of all the valleys to avoid creating uneven pressure that could throw the ridge boards and valley rafters out of alignment.

I nail the tops of the jack rafters to the ridge board first, holding the bottoms in position along the valley rafter. When I nail the bottoms, I hold the top edge of the cheek cut above the top edge of the valley rafter (see the left photo on p. 87). The tops of the jack rafters must be higher than the top of the valley rafter so that the plywood sheathing meets in the middle of the valley, preventing a flat spot in the valley.

Jack Rafters

5. CUT JACK RAFTERS IN PAIRS
Both cuts on the jack rafters are marked plumb for the 12-in-12 pitch of the roof, but the cheek cut, where the jack rafter meets the valley rafter, is beveled 45 degrees. The length of the jack rafters given by the calculator is the theoretical length measured to the center of the valley rafter, so the jacks need to be shortened. To calculate the theoretical lengths of the jack rafters from shortest to longest, use the on-center spacing (16 in., 32 in., etc.) as the run. Then subtract half the diagonal thickness of the valley rafter to get the actual length.

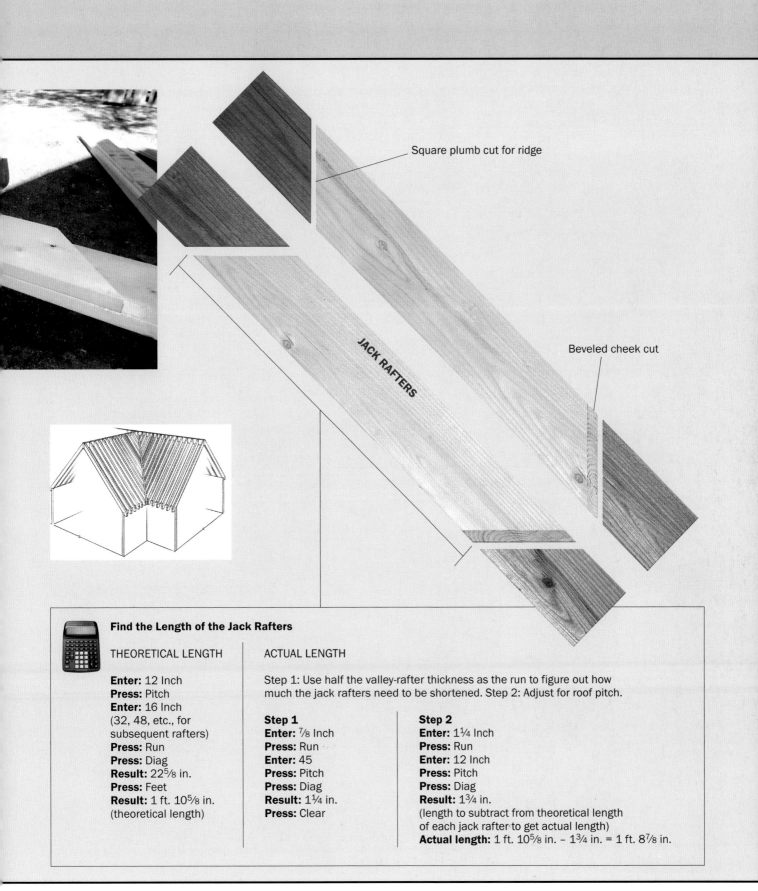

Square plumb cut for ridge

JACK RAFTERS

Beveled cheek cut

Find the Length of the Jack Rafters

THEORETICAL LENGTH

Enter: 12 Inch
Press: Pitch
Enter: 16 Inch
(32, 48, etc., for subsequent rafters)
Press: Run
Press: Diag
Result: $22^5/8$ in.
Press: Feet
Result: 1 ft. $10^5/8$ in. (theoretical length)

ACTUAL LENGTH

Step 1: Use half the valley-rafter thickness as the run to figure out how much the jack rafters need to be shortened. Step 2: Adjust for roof pitch.

Step 1
Enter: $^7/8$ Inch
Press: Run
Enter: 45
Press: Pitch
Press: Diag
Result: $1^1/4$ in.
Press: Clear

Step 2
Enter: $1^1/4$ Inch
Press: Run
Enter: 12 Inch
Press: Pitch
Press: Diag
Result: $1^3/4$ in.
(length to subtract from theoretical length of each jack rafter to get actual length)
Actual length: 1 ft. $10^5/8$ in. – $1^3/4$ in. = 1 ft. $8^7/8$ in.

6. LAY OUT JACK-RAFTER SPACING

The key to laying out the jack rafters along the ridge boards is to begin the layout at the center of their intersection. On the valley rafter, the on-center spacing needs to be adjusted for pitch. Use 16-in. o.c. spacing as the run to do the calculations (see below).

Find Jack-Rafter Spacing
Enter: 16 Inch
Press: Run
Enter: 12 Inch
Press: Pitch
Press: Hip/V
Result: $27\frac{11}{16}$ in.
(This measurement is the location of the first jack rafter, measured from the intersection of the ridges, and it's the on-center spacing for subsequent jacks.)

Install the jacks high. Raise the jack rafters above the edge of the valley rafter so that the roof sheathing will meet in the middle. A scrap of wood makes an ideal gauge to make sure each jack is installed the same.

Why Use a Construction Calculator?

A construction calculator simplifies the math associated with building and carpentry projects. It allows users to work in feet and inches, and has keys labeled with common building terms such as rise and run, and the names of framing members, such as valley rafter and jack rafter, rather than mathematical terms. Follow the examples throughout the chapter to see how a calculator is used to frame a valley roof. Construction calculators are available at most hardware stores, or from Calculated Industries (800-854-8075; www.calculated.com).

To figure out how high to hold the jack rafter, I put a straightedge on the top of the jack rafter and adjust it until the straightedge hits in the center of the valley rafter. Then I nail the jack in place and measure its height above the valley rafter. Once the first jack rafter is installed, I make a gauge the same thickness as the height difference and use it for the rest of the jacks. I install the rest of the jack rafters from the bottom of the valley. Working from the bottom allows me to create a ladder with the jacks, which I can climb as I install consecutively higher rafters. The work's not over yet, though. Shingling a valley also can be a brain teaser. Visit www.finehomebuilding.com to read "Four Ways to Shingle a Valley."

Fine Homebuilding *contributing editor* **Rick Arnold** *is a builder in Wickford, Rhode Island. Photos by Brian Pontolilo.*

A Different Approach to Rafter Layout

■ BY JOHN CARROLL

Earlier this year, I helped my friend Steve build a 12-ft. by 16-ft. addition to his house. A few days before we got to the roof frame, I arrived at his place with a rafter jig that I'd made on a previous job. I'm a real believer in the efficiency of this jig, so I told Steve that it would enable me to lay out the rafters for his addition in 10 minutes.

His look suggested that I had already fallen off one too many roofs. "Come on, John," he said. "Ten minutes?" I bet him a six-pack of imported beer, winner's choice, that I could do it.

Like most builders, I have a long and painful history of underestimating the time different jobs require. In this case, however, I was so certain that I agreed to all of Steve's conditions. In the allotted time, I would measure the span of the addition; calculate the exact height that the ridge should be set; measure and mark the plumb cut and the bird's mouth on the first rafter; and lay out the tail of the rafter to shape the eaves.

When the moment of reckoning arrived, we set a watch, and I went to work. Eight minutes later, I was done and in the process

secured the easiest six bottles of Bass Ale® in my life. With this layout in hand, we framed the roof in 5½ hours.

Why Should It Take an Hour to Do a 10-Minute Job?

That evening, as we enjoyed my beer, Steve's wife asked him how long he would have taken to do the same layout. "An hour," he said, "at least." Steve is a seasoned carpenter who now earns a living as a designer and construction manager. So why does a 10-minute job require 60, or possibly 90, minutes of his time? The answer is that Steve, like many builders, is confused by the process.

The first framing crew I worked with simply scaled the elevation of the ridge from

Laying out rafters doesn't need to be complicated or time-consuming. Using a jig makes the process quick and easy.

Rafter jig doubles as a cutting guide. Scaled to the 12-in-12 roof pitch and with a 1x3 fence nailed to both sides of one edge, this plywood jig is used to lay out plumb and level cuts on the rafters. Notice that the fence is cut short to make room for the circular saw to pass by when the jig is used as a cutting guide. The jig will be used later to lay out plywood for the gable-end sheathing.

If you've always been vexed by roof framing, you may find my way easier to understand than most.

the blueprint and then installed the ridge at that height. Once the ridge was set, they held the rafter board so that it ran past both the ridge and the top plate of the wall, then scribed the top and bottom cuts. Then they used this first rafter as a pattern for the rest. This technique worked. And because it's so simple and graphic, I'm convinced that it still is a widespread practice.

There are several reasons why I retired this method decades ago. To begin with, I haven't always had a drawing with an elevation of the roof system, which means that I couldn't always scale the height of the ridge. Second, it's just about impossible to scale the ridge with any degree of precision. Because of this fact, these roofs usually end up merely close to the desired pitch. Third, this method typically leaves the layout and cutting of the rafter tail for later, after the rafters are installed.

My technique is also different from the traditional approach espoused in most carpentry textbooks, which I've always found to be obscure and confusing. In rafter-length manuals and in booklets that come with rafter squares, dimensions are generally given in feet, inches, and fractions of inches. I use inches only and convert to decimals for my calculations. To convert a fraction to a decimal, divide the numerator (the top number) by the denominator (the bottom number). To convert a decimal to sixteenths, multiply the decimal by 16 and round to the nearest whole number. That number is the number of sixteenths.

Another difference in my approach is the measuring line I use. As the drawing on the facing page shows, the measuring line I employ runs along the bottom edge of the rafter. In contrast, most rafter-length manuals use a theoretical measuring line that runs from the top outside corner of the bearing wall to a point in the center of the ridge. A final thing that I do differently is use a site-built jig instead of a square to lay out the cuts on the rafter.

There are lots of ways to lay out rafters, but if you've always been vexed by roof framing, I think you'll find my way easier to understand than most.

Run and Roof Pitch Determine the Measuring Triangle

Framing a roof can be a little intimidating. Not only are you leaving behind the simple and familiar rectangle of the building, but you're starting a job where there is a disconcerting lack of tangible surfaces to measure from and mark on. Most of this job is done in midair. So where do you start?

There are only two things you need to lay out a gable roof— a choice of pitch and a measurement.

Use the Measuring Triangle to Find the Rafter Length

1. Find the base of the measuring triangle (the run of the roof). Measure between bearing walls and subtract width of ridge, then divide by 2. (271 – 1.75 = 269.25; 269.25 ÷ 2 = 134.63).

2. Find the altitude of the measuring triangle (ridge height). Divide the base of the measuring triangle by 12 and multiply the result by the rise of the roof pitch. For a 12-in-12 pitch, the base and the altitude are the same. (134.63 in. ÷ 12 = 11.22; 11.22 x 12 = 134.63). (Editor's note: To present a set of consistent figures, we rounded to 134.63.)

3. Find the hypotenuse (rafter length). Divide the base of the triangle by 12 and multiply the result by the hypotenuse of the roof pitch, which is listed as length of common rafter on the rafter square (see the photo on p. 93). (134.63 in. ÷ 12 = 11.22; 11.22 x 16.97 = 190.40).

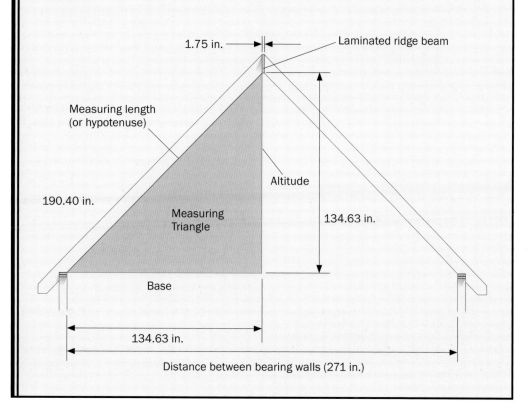

1.75 in.

Laminated ridge beam

Measuring length (or hypotenuse)

Altitude

190.40 in.

Measuring Triangle

134.63 in.

Base

134.63 in.

Distance between bearing walls (271 in.)

There are only two things you need to lay out a gable roof. One is a choice of pitch, and the other is a measurement. Usually the choice of pitch was made before the foundation was poured. The measurement is the distance between the bearing walls (see the drawing on p. 91).

After taking this measurement, deducting the thickness of the ridge, and dividing the remainder in half, you have the key dimension for laying out the rafters. This dimension could be called the "run" of the rafter; but because it is slightly different from what is called the run in traditional rafter layout, I'll use a different term. I'll call it the base of the measuring triangle. The measuring triangle is a concept that I use to calculate both the correct height of the ridge and the proper distance between the top and bottom cuts of the rafter, which I call the measuring length.

Working with this measuring triangle takes a little getting used to. The biggest problem is that when you start the roof layout, only one-third of the measuring triangle exists. As we've just seen, you find the base of the measuring triangle by measuring existing conditions. Then you create the altitude and the hypotenuse (same as the measuring length) by using that base and some simple arithmetic.

To see how this works, let's look at the steps I followed to frame the roof of a 12-ft. by 24-ft. addition I recently finished.

Build the Rafter Jig First

While the rest of the crew finished nailing down the second-floor decking, I began fabricating a rafter jig based on the pitch of the roof (see the left photo on p. 90). The desired pitch was 12-in-12 (the roof rises 12 in. vertically for every 12 in. of run horizontally). The basic design of this jig was simple, and the cost was reasonable: three scraps of wood and 10 minutes of time.

The first step in making this jig was finding a scrap of plywood about 30 in. wide, preferably with a factory-cut corner. Next, I measured and marked 24 in. out and 24 in. up from the corner to form a triangle. After connecting these marks with a straight line, I made a second line, parallel to and about 2½ in. above the first—the 1x3 fence goes between the lines—then cut the triangular-shaped piece along this second line. To finish the jig, I attached a 1x3 fence on both sides of the plywood between these two lines. I cut the fence short so that it didn't run all the way to the top on one side. That allowed me also to use the plywood as a cutting jig running my circular saw along one edge without the motor hitting the fence and offsetting the cut (see the right photo on p. 90).

Here I should pause and note an important principle. The jig was based on a 12-in-12 pitch, but because I wanted a larger jig, I simply multiplied both the rise and run figures by the same number—two—to get the 24-in. measurement. This way, I enlarged the jig without changing the pitch. This principle holds true for all triangles. Multiply all three sides by the same number to enlarge any triangle without changing its proportions, its shape, or its angles.

To use this jig, I hold the fence against the top edge of the rafter and scribe along the vertical edge of the plywood jig to mark plumb lines and along the horizontal edge to mark level lines.

There are at least four reasons why I go to the trouble to make this jig. First, I find it easier to visualize the cuts with the jig than with any of the manufactured squares made for this purpose. Second, identical layouts for both the top (ridge) and bottom (eave) cuts can be made in rapid succession. Third, I use the plumb edge as a cutting guide for my circular saw. Finally, I use the jig again when I'm framing and sheathing the gable end, finishing the eaves and rake, and installing siding on the gable. I also save the jig for future projects.

Step One: Determine the Base of the Measuring Triangle

The base of the measuring triangle is the key dimension for roof layout. In my system, the base of the triangle extends from the inside edge of the bearing wall to a point directly below the face of the ridge. In this addition, the distance between the bearing walls was 271 in. So to get the base of the measuring triangle, I subtracted the thickness of the ridge from 271 in. and divided the remainder by 2. Because the ridge was 1¾-in.-thick laminated beam, the base of the measuring triangle turned out to be 134.63 in. (271–1.75 = 269.25; 269.25 ÷ 2 = 134.63).

Step Two: Determine the Altitude of the Measuring Triangle

With the base of the measuring triangle in hand, it was easy to determine both the altitude and the hypotenuse. The altitude of this measuring triangle was, in fact, too easy to be useful as an example. Because we wanted to build a roof with a 12-in-12 pitch, the altitude had to be the same number as the base, or 134.63 in.

Let's pretend for a moment that I wanted a slightly steeper roof, one that had a 14-in-12 pitch. In a 14-in-12 roof, there are 14 in. of altitude for every 12 in. of base. To get the altitude of the measuring triangle, then, I would find out how many 12-in. increments there are in the base, then multiply that number by 14. In other words, divide 134.63 by 12, then multiply the result by 14. Here's what the math would look like: 134.63 in. ÷ 12 = 11.22; 11.22 x 14 = 157.08 in. Wouldn't it be nice if finding the hypotenuse of the measuring triangle was so simple? It is.

Step Three: Determine the Hypotenuse of the Measuring Triangle

Now let's return to our 12-in-12 roof. The base and the altitude of the measuring triangle are both 134.63 in. But what's the hypotenuse? One way to solve the problem of finding the hypotenuse is to use the Pythagorean theorem: $A^2 + B^2 = C^2$ (where A and B are the legs of the triangle and C is the hypotenuse).

There are other ways to solve this problem—with a construction calculator, with rafter manuals, with trigonometry—but I usually use the principle mentioned in step one. According to this principle, you can expand a triangle without changing the angles by multiplying all three sides by the same number.

> *The base of the measuring triangle is the key dimension for roof layout.*

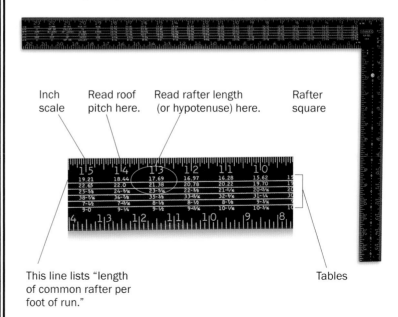

Rafter Square Gives You the Hypotenuse

Find the number for your roof pitch on the inch scale (12 in. for an 12-in-12 pitch). The first number under that is the hypotenuse of a right triangle with a 12-in. base (or run) and an 12-in. altitude (or rise).

Inch scale

Read roof pitch here.

Read rafter length (or hypotenuse) here.

Rafter square

This line lists "length of common rafter per foot of run."

Tables

For more than a century, carpenters have used the rafter tables stamped on rafter squares. The common table shows the basic proportions of triangles for 17 different pitches (see the photo on p. 93).

The base of all these triangles is 12; the altitude is represented by the number in the inch scale above the table. And the hypotenuse is the entry in the table. Under the number 12, for example, the entry is 16.97. This is the hypotenuse of a right triangle with a base and an altitude of 12.

To use this information to create the larger measuring triangle I needed for this roof, I simply multiplied the altitude and the hypotenuse by 11.22. This, you may recall, was the number I obtained in step two when I divided the base, 134.63 in., by 12. Now multiply the altitude and the hypotenuse of the small triangle by 11.22: 11.22 x 12 gives us an altitude of 134.63 in.; and 11.22 x 16.97 gives us a hypotenuse of 190.40 in.

So here is the technique I use for any gable roof. I find the base of the measuring triangle, divide it by 12, and multiply the result by the rise of the pitch to get the altitude of the measuring triangle (which determines the height to the bottom of the ridge). To get the hypotenuse, I divide the base of the triangle by 12 and multiply that by the hypotenuse of the pitch, which is found in the common rafter table.

Say the roof has an 8-in-12 pitch with a base of 134.63. I divide that number by 12 to get 11.22, then multiply that by 8 to get the ridge height of 89.76 in. To get the length of the rafter, I multiply 11.22 by 14.42 (the number found under the 8-in. notation on the rafter table), for a length between ridge and bearing wall of 161.79 in.

The only time I waver from this routine is when a bird's mouth cut the full depth of the wall leaves too little wood to support the eaves. How little is too little depends on the width of the rafter and the depth of the eave overhang, but I generally like to have at least 3 in. of uncut rafter running over the bird's mouth. If I have too little wood, I let the bottom edge of the rafter land on top of the wall rather than aligning with the wall's inside edge. Then I determine how far the rafter will sit out from the inside edge of the bearing wall and use that inside point as the start of my measuring triangle.

Step Four: Setting the Ridge

I determined that the altitude of the measuring triangle was 134.63 in. This meant that the correct height to the bottom of the ridge was 134⅝ in. above the top plate of the wall. (Note: I usually hold the ridge board flush to the bottom of the rafter's plumb cut rather than the top.) To set the ridge at this height, we cut two posts, centered them between the bearing walls, and braced them plumb. Before we installed the posts, we fastened scraps of wood to them that ran about 10 in. above their tops. Then, when we set the ridge on top of the posts (see the photo on the facing page), we nailed through the scrap into the ridge. We placed one post against the existing house and the other about 10 in. inside the gable end of the addition. This kept it out of the way later when I framed the gable wall.

Getting the ridge perfectly centered is not as important as getting it the right height. If opposing rafters are cut the same length and installed identically, they will center the ridge.

Step Five: Laying Out the Main Part of the Rafter

I calculated a measuring length (or hypotenuse) for the rafter of 190.40 in., then used the jig as a cutting guide and made the plumb cut. I measured 190⅜ in. (converted from 190.40 in.) from the heel of the plumb cut and marked along the bottom of the rafter. Rather than have another carpenter hold the end of the tape, I usually clamp a square

Setting the ridge. Temporary posts are set up to hold the ridge at the right height. The posts are braced 2x4s with a 2x4 scrap nailed to the top that rises 10 in. above the top of the post. The ridge is set on top of the post and nailed to the scrap of wood.

A clamped square serves as an extra pair of hands. It's easy to measure along the bottom of the rafter if you clamp a square to the bottom of the plumb cut and run your tape from there.

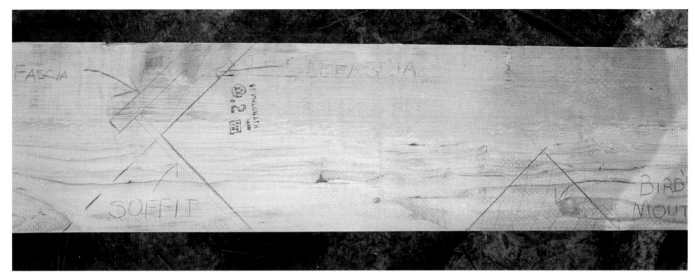

Visualizing the bird's mouth and rafter tail. Red lines on this marked-up rafter show where cuts will be made for the bird's mouth, at right, and rafter tail, at left. To determine the correct cut for the tail, the author marked out the subfascia, soffit, and fascia board.

across the heel of the plumb cut, then pull the tape from the edge of the square (see the photo on the facing page) to determine where the bird's mouth will be.

Once I marked where the bird's mouth would be, I used the jig to mark a level line out from the mark. This would be the heel cut, or the portion of the rafter that sits on top of the bearing wall.

After scribing a level line, I measured in the thickness of the wall, which was 5½ in., and marked. I slid the jig into place and scribed along the plumb edge from the mark to the bottom edge of the rafter. This completed the bird's mouth and, thus, the layout for the main portion of the rafter.

Step Six: Lay Out the Rafter Tail

The eaves on the existing house measure 16 in. out from the exterior wall, which meant I would make the eaves on the addition 16 in. wide. To lay out the rafter tail, I started with the finished dimension of 16 in. and then drew in the parts of the structure as I envisioned it (see the photo above). In this way, I worked my way back to the correct rafter-tail layout.

I began by holding the jig even with the plumb line of the bird's mouth. With the jig in this position, I measured and marked 16 in. in from the corner along the level edge. Then I slid the jig down to this mark and scribed a vertical line. This line represented the outside of the fascia. Next I drew in a 1x6 fascia and a 2x subfascia. I also drew in the ⅜-in. soffit I would use. This showed me where to make the level line on the bottom of the rafter tail.

Step Seven: Preserve the Layout

The only dimension for this layout that I had to remember was 190⅜ in., the hypotenuse of the triangle and the measuring length of the rafter. I wrote the number where I could see it as I worked. To preserve the other three critical dimensions—one for the plumb cut of the bird's mouth and the other two for the rafter tail—I used the rafter jig to extend reference points to the bottom edge of the rafter; then I transferred these marks to a strip of wood, or measuring stick (see the photo on p. 98). I was ready to begin cutting the rafters.

Marking the bird's mouth and tail. Once the start of the heel cut was determined by measuring from the bottom of the plumb cut, this measuring stick was used to transfer the dimensions of the bird's mouth and rafter tail. The cutting lines were marked using the rafter jig.

A rafter jig does the hard work. Once the location of the bird's mouth is determined, the rafter jig is used to mark the level and plumb cuts of the bird's mouth and rafter tail.

Step Eight: Marking and Cutting the Rafters

Some carpenters lay out and cut one rafter, then use it as a pattern for the rest, and I'll often do that on a smaller roof. On this roof, where the rafters were made of 20-ft. long 2x12s and where I was laying them out by myself, this method would have meant a lot of heavy, awkward, unnecessary work. Instead of using a 100-lb. rafter as a template, I used my jig, my tape measure, and the measuring stick.

Moving to the end of the 2x12, I clamped the jig in place and made the plumb cut. (The steep pitch of this roof made clamping the jig a good idea. Usually, I just hold it to the rafter's edge the way you would when using a framing square as a cutting guide.) Then I clamped my square across the heel of that cut, pulled a 190⅜-in. measurement from that point, and marked along the bottom edge of the

board. Next I aligned the first mark on the measuring stick with the 190⅜-in. mark and transferred the other three marks on the measuring stick to the bottom edge of the rafter (see the photo above left).

To finish the layout, I used the jig to mark the level and plumb lines of the bird's mouth and the rafter tail (see the photo above right). For all four of these lines, I kept the jig in the same position and simply slid it up or down the rafter until either the plumb or level edge engaged the reference mark. It was quick and easy.

The two cuts that formed the rafter tail were simple, straight cuts that I made with my circular saw. To cut the bird's mouth, I cut as far as I could with my circular saw without overcutting, then finished the cut with my jigsaw.

John Carroll is a contributing editor to Fine Homebuilding *and the author of* Measuring, Marking and Layout: A Guide for Builders and Working Alone, *published by The Taunton Press.*

Aligning Eaves on Irregularly Pitched Roofs

■ BY SCOTT MCBRIDE

Pick up a catalog of stock house plans in any supermarket these days, and you'll see that cut-up roofs—roofs with lots of hips and valleys—are back in fashion. The highly competitive new-home market has compelled builders to spice up their roofs with tasty devices such as Dutch hips and wall dormers. The desired effect is curb appeal, which is the elusive but all-important quality that plays to the homebuyer's romantic notion of what a dream house should look like.

As long as all intersecting roof slopes are inclined at the same pitch, framing a cut-up roof can be fairly straightforward: When the slopes are the same, all of the hips and the valleys run at 45 degrees in plan. Consequently, all hip-, valley- and jack-rafter cuts

Spicing up a gable complicates framing and trimming. A pair of steeply pitched gables adds curb appeal to this house with a medium-pitch gable roof. Getting the rafter tails to line up required raising the wall plates on the bays, or kickouts.

can be made on a simple 45-degree bevel (the cheek cuts), and only two plumb-cut angles are required: the common-rafter plumb cut and the hip/valley-rafter plumb cut.

However, combining steep-roofed projections with a medium-pitch main roof is a good way to compromise between cost and curb appeal.

A roof system usually starts with a main gable, and increasing the gable's pitch dramatically increases material and labor costs. Cosmetic roof features such as dormers are much smaller, so increasing their pitch won't have the same impact on cost as will increasing the pitch of the main roof.

Because the usual purpose of cosmetic features is to lend drama to a home's facade, there may be a strong incentive to make secondary roofs steeper than the main roof, especially on the street side.

While adding steeply pitched features to a medium-pitch main roof might seem like an ideal way to increase curb appeal, it complicates the framing considerably. I recently built a house that has such an unequal-pitch condition, and here I want to talk about some of the difficulties I encountered and how I resolved them.

The Particulars of This Job

The house is rectangular in plan, except for two rectangular bays, or kickouts, extending 16 in. beyond the front wall (see the photo on p. 99). The kickouts are topped with 12-in-12 gable roofs. The main roof of the house has a 7-in-12 pitch. Because of the different roof pitches, the valleys don't run at 45 degrees in plan; they're angled toward the lower-pitch main roof. I had to figure out what that angle was and then how to frame the valley.

The eaves were to overhang 12 in. on both the main roof and the kickout gables, and a sloping soffit was to be nailed directly to the rafter tails. If I built the main wall and the kickout walls the same height, the kick-

out rafter tails would end 5 in. lower than the rafter tails on the main roof (7-in-12 vs. 12-in-12). That would misalign the fascia boards and the sloping soffits.

Complicating matters further, the rafters for the kickout gables were to be 2x6, while the main-roof rafters were to be 2x8. To get a grip on all of these variables, I headed to the drawing board.

Drawing the Cornice Section

I always begin roof framing with a full-scale cornice section drawn on a piece of plywood or drywall (see the drawing on the facing page). In this case I drew one cornice section on top of the other—the 7-in-12 main roof cornice section and the 12-in-12 kickout roof cornice section. The superposed drawing provided me with the length of the rafter tails, the location of the bird's mouths, the width of the fascia, and the depth of the sloping soffits. From the drawing, I also determined how high I'd have to raise the kickout wall so that the kickout fascia would line up with the main fascia.

I began by drawing in the 7-in-12 overhang for the main roof, with the lower edge of the 2x8 rafter starting at the inside corner of the wall plate, the typical location for a bird's mouth. Underneath the 2x8 tail I drew the main-roof soffit. From the point where the face of the soffit meets the back of the fascia, I drew a line at a 12-in-12 pitch to represent the more steeply pitched kickout soffit. Next, I drew a parallel 12-in-12 line from the top end of the 2x8 tail of the main roof; this line represented the top edge of the kickout rafter tail. I now had the top and the bottom edges of my rafter tails aligned at a point 11¼ in. away from the wall. (The ¾-in. thickness of the fascia would increase the overhang to 12 in.)

Next, I drew the 5½-in. width of the 2x6 kickout rafter. The remaining distance between the lower edge of the 2x6 and the back of the kickout soffit is made up by

Superposing Main-Roof and Kickout-Roof Cornices

This drawing shows how the wall plate was built up to catch the kickout rafters, which made eaves alignment possible. Starting from the top inside edge of the main wall plate, the author drew the main rafter at a 7-in-12 pitch, then added the 2x6 kickout rafter at 12-in-12 in line with the top end of the main-rafter tail. The bottom of the 2x6 rafters required shims so that the soffits could line up.

2x6 kickout rafter

2x8 main rafter

Kickout bird's mouth

Added plates

Fascia

Standard double-top plate

Shim

Main soffit

Kickout soffit

shimming. The shims cover the underside of the 2x6 kickout tails and extend up along the lower edge of the kickout barge rafters to keep the eave soffit flush with the rake soffit.

At this point I could see roughly how much I needed to raise the kickout wall plate. The exact elevation of the kickout bird's mouth wasn't critical because it's in the attic above the second-story ceiling. I built up layers of 2x4 until the raised plate gave good bearing for the kickout rafters (see the photo at right and the drawing above). With the kickout rafters sitting higher on the wall than the main-roof rafters, the fascia and the sloping soffit could flow in a smooth line from one roof to the other.

Additional top plates provide bearing for the kickout's 2x6 rafters while picking them up enough to line up the fascia boards.

Simplifying Valley Construction

There are two methods of building valleys. The first, called a framed valley, employs a valley rafter that supports jack rafters coming down from both intersecting roof surfaces.

A simpler approach, known as a California roof or a farmer's valley, is to build the main roof all the way across and frame the intersecting roof on top of it. Instead of valley rafters, you nail valley boards flat on the main roof and frame jack rafters for the smaller roof only (see the photo below).

Because a California roof typically sits on the main-roof sheathing, it can't be used if the smaller roof will have a cathedral ceiling. But there were no cathedral ceilings in this house, so I could use the California approach to frame this roof.

Building a California roof simplified the framing significantly because an unequal-pitch valley has several peculiar traits. First,

given that the overhang is the same for both roofs, the valley will not cross over the inside corner where the walls intersect, as is usually the case. Rather, it will veer toward the roof with the lower pitch. A valley rafter's location would have to be figured out beforehand from studying a plan view of the roof framing.

Furthermore, an unequal-pitch hip or valley rafter requires two different edge bevels at the point where it hangs on intersecting ridges or headers. One of the bevels will be sharper than 45 degrees, so it cannot be cut with a standard circular saw. With the California roof, I avoided the problem of dissimilar edge bevels and the hassle of locating valley rafters.

I located the off-center valley by snapping lines on the main-roof sheathing. First, I installed the kickout ridge and its common rafters. Then, at the peak of the kickout gable, I anchored a chalkline, stretched it diagonally across the kickout

One roof framed on another. A California roof features valley boards that lie flat on the main roof sheathing. As opposed to a valley rafter, there's only one set of jack rafters to cut and one cheek-cut setting on the circular saw.

common rafters, and marked a point somewhere near the eaves where the chalkline hit the main-roof sheathing. The top of the valley occurs where the kickout ridge dies into the main roof, so, by striking a line across these two points, I located the valley boards. I beveled the edges of the valley boards and nailed them flat on the main-roof sheathing.

Mitering the Soffits

Once I had the kickout jack rafters and sheathing in place, I turned my attention to the cornice. Thanks to the raised plate on the kickout wall, all of the rafter tails lined up, and the fascia flowed smoothly around the corner. But now I had to make the sloping soffits do the same thing.

The main question was, at what angle should the soffit boards be cut to create a clean miter at the inside corner (where the main wall and the kickout wall intersect)?

I could cut some scrap pieces with 45-degree face cuts and continue adjusting the angle by trial and error until I had a good fit. Or I could stay on the ground, figure out the angles on paper, and install the soffits on the first try.

I opted for method two, and I accomplished this through graphic development. Graphic development is a way of taking a triangle that occurs in space, such as the gable end of a roof, and pushing it down on a flat surface where it can be measured accurately. All you need is a pencil, some paper, a framing square, and a compass. A stubborn disposition helps, too.

To figure out the miter angles of the soffits, I began by drawing a plan view of the wall lines and the fascia lines (see the drawing on p. 104). I then drew in plan views of the main-roof rafter tail and the kickout rafter tail, each perpendicular to its respective wall. In addition, I drew elevation views of the same rafter tails. I used the numbers 7 and 12 on the framing square to draw the main-rafter elevation view and 12 and 12 for

the kickout elevation view. As the plan view and the elevation view of each tail cross the wall lines, they show the vertical rise of the tails: 7 in. for the main roof, 12 in. for the kickout.

Next, I needed to draw the valley. Remember, it doesn't run at 45 degrees. I already had the end of the valley: the point where the fascias intersected. What I needed was another point farther up the valley so that I could draw the valley line. Because I already had drawn the 12-in. kickout elevation view, I decided to find the point where the valley rises 12 in.

First, I extended the kickout-wall line in the direction of the valley. At this line, the kickout roof rises 12 in. above the fascia, so the line represents plan views of both the kickout wall and the kickout roof's 12-in. rise line. Then, I extended the main rafter tail and drew a perpendicular line showing the plan view of the main roof where it rises 12 in. above the fascia. The intersection of both of the 12-in. plan-view rise lines is a point on the valley, and, by connecting this point with the inside corner of the fascias, I drew the valley in plan.

To determine the miter angles (or face cuts) for the soffit material, I used a compass to swing the 12-in.-high elevation view of each common rafter tail directly over the plan view. This point represents the actual length of the rafter where it rises 12 in. (as opposed to the foreshortened length of the rafter when seen in plan).

Then, I drew lines perpendicular to the plan views of the rafters at the points where my compass intersected them. These are labeled elevation lines on the drawing. Next, I intersected the plan line of one soffit with the elevation line of the other soffit. I connected these intersecting points to the inside corner of the fascia, giving me the angle of each soffit's face cut.

Try to imagine the inside soffit edges rising up while the outside soffit edges remain "hinged" along the fascia line. When the inside edges have risen 12 in., the soffits touch

Combining Plan Views and Elevations

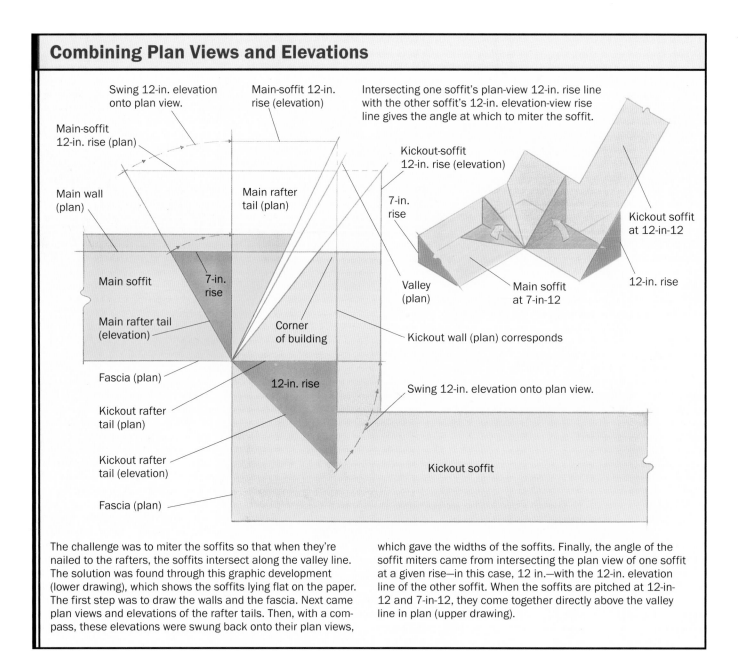

Swing 12-in. elevation onto plan view.

Main-soffit 12-in. rise (plan)

Main-soffit 12-in. rise (elevation)

Intersecting one soffit's plan-view 12-in. rise line with the other soffit's 12-in. elevation-view rise line gives the angle at which to miter the soffit.

Main wall (plan)

Main rafter tail (plan)

Kickout-soffit 12-in. rise (elevation)

7-in. rise

Main soffit

7-in. rise

Kickout soffit at 12-in-12

Main rafter tail (elevation)

Corner of building

Valley (plan)

Main soffit at 7-in-12

12-in. rise

Fascia (plan)

Kickout wall (plan) corresponds

Kickout rafter tail (plan)

12-in. rise

Swing 12-in. elevation onto plan view.

Kickout rafter tail (elevation)

Kickout soffit

Fascia (plan)

The challenge was to miter the soffits so that when they're nailed to the rafters, the soffits intersect along the valley line. The solution was found through this graphic development (lower drawing), which shows the soffits lying flat on the paper. The first step was to draw the walls and the fascia. Next came plan views and elevations of the rafter tails. Then, with a compass, these elevations were swung back onto their plan views, which gave the widths of the soffits. Finally, the angle of the soffit miters came from intersecting the plan view of one soffit at a given rise—in this case, 12 in.—with the 12-in. elevation line of the other soffit. When the soffits are pitched at 12-in-12 and 7-in-12, they come together directly above the valley line in plan (upper drawing).

along their face cuts, forming a valley, or as viewed from below, an upside-down hip (see the photo on the facing page). The meeting takes place directly over the plan view of the valley line.

The face cut for the kickout soffit was the same as the face cut for the kickout roof sheathing, which happened to be 1x6 but could just as easily have been 4x8 sheets. The sheathing sits on top of the rafters, and the soffit hangs below. Otherwise, they're the same.

Because of its steeper pitch, the kickout soffit dies square into the house for a short distance before mitering with the main-roof soffit. I could tell from the graphic development where to cut the kickout soffit along the main-wall line. The pieces fit on the first try.

Joining Mitered Soffits

The edge bevel for the soffits wasn't critical because the back side doesn't show. I just

Sloping soffits make for tricky miters. One problem with this roof was getting the sloping soffits of two different roof pitches to flow smoothly around the inside corner. The miter mirrors the offset valley, so the kickout soffit dies square into the main wall. A beveled 2x4 provides backing along the miter joint.

cut them at 45 degrees, which undercut the pieces more than necessary and assured a tight miter. Fastening the intersecting soffits presented a problem, however, because I didn't have a valley tail to nail the ends of the soffit plywood to. (That would have been the only good reason for using a framed valley here instead of a California valley.)

To solve the problem, I connected the main-roof soffit and the kickout soffit along their intersection with a beveled 2x4 backerboard. I beveled the 2x4 to match the angle

of the valley trough. The 2x4 backerboard didn't have to fit tightly between the fascia boards and the house because the ⅜-in. soffit material was pretty stiff. Instead, I just cut the ends of the backerboard for a loose fit and pulled it against the soffits with galvanized screws.

Scott McBride is a contributing editor of Fine Homebuilding. *His book* Build Like a Pro: Windows and Doors *is available from The Taunton Press. McBride has been a building contractor since 1974.*

Vinyl Siding Done Right

■ BY MIKE GUERTIN

I cringed the first time a custom-home client asked for vinyl siding on a new $400,000 house. But I had little choice. Our market was dead. So I took the project. Now more than half of my custom-home clients ask for vinyl, and I actually like installing it. Although vinyl's installation is faster, making vinyl look good on a house is more challenging than wood or fiber cement.

Hands on is always the best way to learn, but you can get basic installation instructions from the Vinyl Siding Institute, and manufacturers offer brand-specific installation manuals as well. Here, I highlight a few techniques for enhancing the appearance of clapboard-style vinyl siding in new construction.

A word of caution: Vinyl siding leaks a lot. Water enters where vinyl panels overlap and at the end gaps inside the trim's receiving channels. Manufacturers even punch weep holes into the bottom edge of siding to let the water escape. Besides being careful to maintain the integrity of the tar paper or housewrap behind vinyl, I recommend flashing windows and doors to direct water back out when it gets past the flanges or casing.

Begin with the Trim

Vinyl siding depends on a few special trim pieces that either hold everything together or hide the edges and ends of the panels. Undersill trim, for instance, is installed under windowsills or other horizontal projections and grasps the top edge of the siding panel below it (see top right photo on p. 113). The J-channel receives the ends of siding panels where they abut windows, doors, or other walls. Most pieces of vinyl trim, though, go by the same names as their wood-siding counterparts: soffits, fascia, and corner boards.

I start by installing the high trim (rakes, soffits, and fascias). For rakes, I usually bend aluminum coil stock on a sheet-metal brake to wrap the face and bottom edge of a 1x rake board. Manufacturers also offer vinyl-coated aluminum and all-vinyl trim stock that are bendable and give excellent results. The rake board is furred out with a board narrower than the rake to create a space for the siding ends—sort of a site-made J-channel. Prep for soffit installation begins as I frame the house. Along the tails of the

Pushing up locks panels together. The bottom edge, or butt, of one panel clicks into the locking channel at the top of the preceding course, and the siding goes up.

The author cuts soffit panels ¼ in. shorter than the soffit depth to allow for thermal expansion; he then traps the soffit panels with site-bent aluminum fascia trim and vinyl window casing. The window casing captures the top of the siding panels and mimics a frieze board, which looks better than narrow J-channel.

Site-bent aluminum fascia captures front of soffit panels.

Window and door casing supports soffit panels at wall.

Sources

Vinyl Siding Institute
888-367-8741
www.vinylsiding.org

rafters or trusses, I install a 2x subfascia, then hang the soffit panels and fascia on that (see the photo above). Where soffit panels meet the wall of the house, they are supported by J-channel, undersill trim, or window casing.

Snapping a Line Keeps Corner Boards Straight

After the rakes and soffits are finished, I install the corner boards. Because corner boards are hollow and flexible, nailing them up straight can be difficult. Snapping chalklines on both sides of the corner ensures that the corner board goes on straight (see the photos on p. 110). I use short corner-board sections as templates to mark the wall for the chalkline.

I begin nailing the corner board at the top and use sheet-metal snips to cut the flange back where it abuts the already installed soffit trim. The top two nails (one on each side of the corner) are placed at the top edge of the nail slots and are set tightly. These nails lock the corner-board top in place so that it expands downward rather than upward into the soffit. I drive the rest of the nails (spaced about 12 in. apart) in the middle of the slots to permit that expansion, leaving the head about ¹⁄₁₆ in. proud of the nailing flange so that the vinyl can move.

Aluminum fascia supports soffit panels.

Wide casings replace J-channel.

Hanging Basics

HANG SIDING, DON'T NAIL IT
Unless your brand of vinyl has a design feature that eliminates concerns about nail depth, leave the heads of nails about $\frac{1}{16}$ in. proud of the vinyl to allow sideways movement.

GIVE PANELS ROOM TO MOVE IN THE RECEIVING CHANNELS
When installing at temperatures above 90°F, leave a $\frac{1}{4}$-in. expansion gap to prevent panels from buckling. Leave $\frac{1}{2}$ in. below 30°F and $\frac{3}{8}$ in. at temperatures in between.

MINIMIZE VISIBILITY OF PANEL OVERLAPS
Start the first panel of each course at the end opposite the most common viewing point so that subsequent overlaps face away. Avoid uniform, stair-step installation patterns that catch the eye.

KEEP IT STRAIGHT
Snapping a line ensures straight, even courses. Straight lines are especially important on walls broken up by windows or doors.

AVOID J-CHANNEL
Wide window and door trim and inside corners can replace most J-channel, the hallmark of most bad vinyl jobs.

VINYL LEAKS
A reliable drainage plane and thorough flashing are the weather barrier.

ALIGN PANELS WITH TOPS OF WINDOWS
Calculate the height of starter strip so that the siding's shadowlines meet the tops of the most visible windows.

Receiving channels in trim hide panel ends.

Starter strip can hang below sheathing to align siding panels with tops of windows.

Siding panels overlap to allow expansion.

Alignment Is Crucial

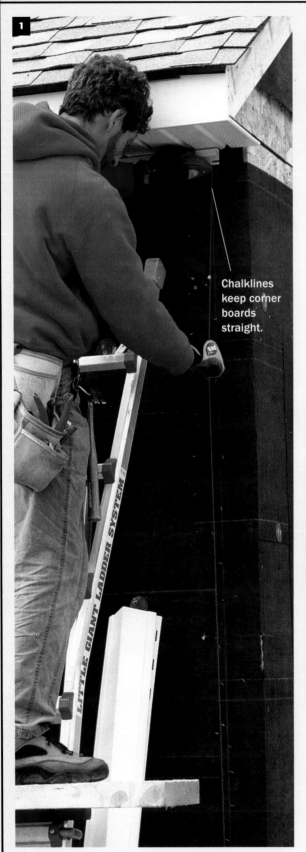

1

Chalklines keep corner boards straight.

Adjusting the starter-strip height can keep panels even with the tops of windows and doors. Here, the author dropped the starter strip below the sheathing by pop-riveting it to pieces of coil stock.

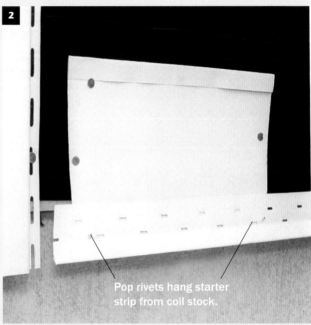

2

Pop rivets hang starter strip from coil stock.

3

Chalklines keep panels straight and align them with tops of windows.

Snapping chalklines for corner boards and siding panels ensures that the flexible vinyl stays straight. Siding panels have some vertical play, so checking overall siding height frequently allows installers to make small adjustments and avoid accumulated error.

After determining exactly where at the bottom of the wall the starter course will begin (see top right photo on facing page), I trim all the corner boards just below the bottom of the starter strip (¼ in. in winter, ½ in. in summer) and finish nailing them in place. Except for window trim, most of the prep work is now done.

Accommodating Thermal Expansion

Vinyl-siding manufacturers try hard to make their panels look like real wood. Although the manufacturers are getting better, vinyl installs, behaves, and performs much differently from the natural material it mimics.

Nailing Details and Expansion Gaps Accommodate Panel Movement

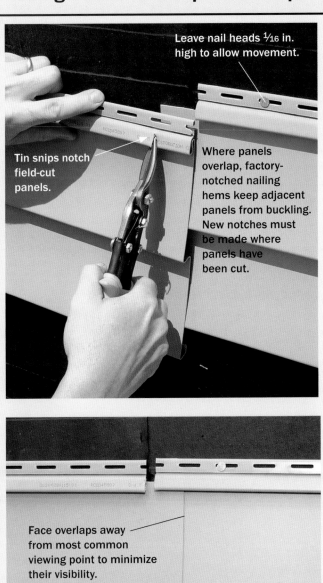

Leave nail heads ¹⁄₁₆ in. high to allow movement.

Tin snips notch field-cut panels.

Where panels overlap, factory-notched nailing hems keep adjacent panels from buckling. New notches must be made where panels have been cut.

Panels overlap so that they can slide past each other as they expand and contract with temperature changes. To allow this movement, the panels are made with the nailing hems notched at the end. This notch must be duplicated on field-cut panels. Most nail heads are left high to avoid trapping the panels. After the hems are locked, the endlap is backed off for no more than 1 in. overlap at the ends of the panels.

Butting panels too tightly into trim's receiver channels can cause buckling in high temperatures.

Face overlaps away from most common viewing point to minimize their visibility.

Fixing the center of a panel with two nails ensures that thermal expansion happens equally at each end.

Whereas wood moves in response to moisture content, vinyl moves in response to temperature changes. And it can move a lot through expansion and contraction, especially along the length of the clapboard-style panels. To compensate for this movement, adjacent siding panels are overlapped so that they can slide sideways past each other without buckling or creating gaps. Nails are left proud of the surface to allow for this movement. Where panels meet windows, doors, or corner boards, they are cut short. Receiving channels on the trim conceal the expansion gap. Trim along roof slopes should be set ½ in. above hot asphalt shingles.

Minimize J-Channel around Openings

Windows and doors need to be surrounded with a receiver channel to accept the ends of the siding. This process usually means wrapping them all with J-channel, one of the dead giveaways that a house is covered in vinyl. Minimizing J-channel takes planning and creativity. When I know a house will be sided in vinyl, I select windows that have either an integral J-channel or a slot to receive an applied cap that creates a receiver channel. Alternatively, I use vinyl window and door casing, which is essentially a 2½-in. wide J-channel (see the photo on p. 109). The greater width can enhance the window rather than detract from it, as a narrow band of standard J-channel does.

Around doors, I use a couple of other tricks to build a receiver channel. Using either a preformed foam plastic or a custom wood door surround is the easiest. I simply apply a 1x3 wood spacer over the door jamb and exterior sheathing (see photo at right on p. 114). Then I fasten the decorative trim over the spacer to create a ¾-in.-wide pocket that receives the ends of the siding. Alternatively, I can trim the door with furring and bend aluminum coil stock around it.

Notching siding into J-channel keeps water outside the wall.

When my bag of tricks runs out and I'm forced to wrap an opening with J-channel, I try to make it look as clean as possible. You can't count on the leg of the J-channel at the top of a window or door to serve as head flashing. When using J-channel, I always install a drip cap that slips beneath the housewrap and laps over the window or door. The J-channel goes over the flashing.

Align Siding Panels with Window Tops and Intersecting Roofs

Perhaps the most fundamental trick to making vinyl look good begins well before installation. One of my goals when planning wood siding is to have the bottom edges of the clapboards line up with the tops and bottoms of windows and doors whenever possible. This alignment unites the exterior look of the building. Alignment is difficult with vinyl because the course exposure isn't adjustable. And this lack of alignment is one of vinyl's telltale signs.

To minimize this problem, I begin planning the vinyl course layout before the foundation of the house is poured. Because the limiting factor is the vinyl itself, I choose the siding for the house first. The three most common vinyl-siding patterns are double four, triple three, and double five. That is, vinyl panels are made to look like two 4-in. clapboards, three 3-in. clapboards or two 5-in. clapboards. Their total exposures are fixed: 8 in., 9 in., and 10 in.

Knowing the exposure allows me to plan at least some of the rest of the house around it. I plan foundation drops, those spots where the foundation steps up or down to keep pace with the grade, on increments to match the siding height. It's easy to position the window rough openings so that the tops of windows match a siding course. The bottoms of windows are hit or miss unless I'm

Window Trim Is Not Flashing

Flashing tape

Tab in top piece laps over lower trim.

Side piece is not mitered at top.

Whether it's J-channel or a wider piece of casing, a tab cut in the top piece of the window trim wraps over the side to divert water. In addition, the metal head flashing tucks behind the tar paper and over the window top or flashing tape is used to seal tar paper to the window's nailing fin.

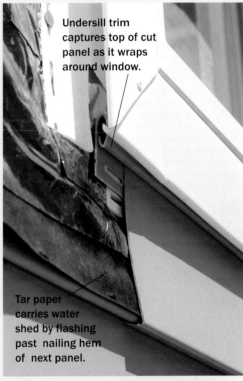

Undersill trim captures top of cut panel as it wraps around window.

Tar paper carries water shed by flashing past nailing hem of next panel.

Behind the window casing, the undersill trim secures the cut vinyl panel under a window. Builder's felt and flashing membrane direct water past the nailing hem of the first full panel below the window, pushing the water outside.

When notching panel at bottom of window, leave ¼-in. expansion space all around.

> *It may seem excessive to plan exterior building elevations to match the siding, but the results are worth the effort.*

using custom-built vinyl windows and can specify window height.

When I have to notch panels at the bottom of windows and at J-boxes (see bottom left photo), I make sure the notch isn't at a joint between panels—both because it's easier to cut one notch than two and because it looks better. I leave a ¼-in. space on the rip beneath the opening and on each side of the window or box to allow for expansion space. I can't adjust the height of the doors, but the width of the surrounding trim can be tweaked to align with the bottom of a siding panel if it's within 1 in. or so.

When porch, garage, or other intersecting roofs interrupt siding, I have to install J-channel to start the panels. But by adjusting the roof pitch just a little so that the butt line matches the roof–wall intersection, I get a clean look. The same goes for the height of a deck. It may seem excessive to plan exterior building elevations to match the siding, but the results are worth the effort.

Mike Guertin is a contributing editor to Fine Homebuilding *magazine and the author of* Precision Framing *and* Asphalt Roofing, *both published by The Taunton Press. He lives in East Greenwich, Rhode Island.*

J-Boxes and Furred-Out Door Trim Replace J-Channel

Trimming around outlets, light fixtures, electric-meter sockets, and ventilation exhaust ports can be accomplished easily with preformed J-boxes. Several manufacturers make special accessory boxes with integral J-channels or snap-on J-surrounds to make trimming the penetration faster, more weather-resistant, and more attractive than something that can be fabricated on site. Whenever possible, I roughly locate the J-boxes on the wall but don't attach them until the siding goes up. This way, I can achieve a cleaner look by precisely locating the top of each box against the bottom of the panel above it.

Around doors, the easiest way to hide panel ends is to fur out the trim so that siding panels abut the furring and the trim forms a receiving channel.

Furring lifts door trim over panel ends.

Utility boxes create receiving channels. A snap-on molding holds panel ends around a light box near the front door.

Door trim replaces J-channel. At a front door, a furring strip lifts the door trim enough to let siding panels tuck in behind.

Installing Horizontal Wood Siding

■ BY FELIX MARTI

I've installed a lot of wood siding during my 20-plus years in the building trades. And whenever I'm given the option, I prefer to install the siding horizontally. On the aesthetic side, I like the way horizontal shadowlines emphasize the shapes of houses. But more important, horizontal siding lasts longer than vertical siding, especially in wet climates. That's because water running down a piece of vertical siding inevitably hangs along the bottom edge for a while, where it can be wicked into the end grain of the wood. The result: rotted wood, peeling paint, or both.

In this chapter I offer some suggestions for selecting wood siding, establishing a workable layout, and then installing the siding. Here in southwestern Colorado, we typically side a house with rough-sawn, unpainted red cedar, such as the channel-rustic siding shown in the photos. But the techniques I'll talk about are equally appropriate for painted clapboards on a colonial house.

The style of your house will strongly influence the siding pattern you choose and the grade of the material. For example, rough-sawn, knotty cedar or redwood evokes a rustic feeling. Clear, vertical-grain clapboards on a crisply painted colonial house, on the other hand, are more patrician. You can bet that patrician costs more than rustic when it comes to siding.

If you are budget-minded and if you plan to paint your siding, there is a relatively new composite product on the market called Inner-Seal® lap siding from Louisiana-Pacific. Inner-Seal is $\frac{7}{16}$ in. thick, and it comes preprimed in 8-in., 9½-in., and 12-in. widths. The pieces are 16 ft. long. I used Inner-Seal siding on a house in the wet part of Oregon with good results. It paints beautifully, and it is stable and straight. I also like the fact that Inner-Seal siding is top-nailed, which conceals the fasteners.

Ordering Material and Checking for Defects

The pattern you choose influences the equation for determining how much siding you need. Measure the surface area of the walls to be covered, then subtract the square foot-

> The style of
> your house will
> strongly influence
> the siding pattern
> you choose and
> the grade of the
> material.

The layout depends on the openings. Full-width siding boards without notches will tuck nicely under the windows when the next two courses are applied. That's because the board at the bottom of the wall was ripped to a narrower width to simplify the cuts around the windows.

Unused finish is returned to its can by opening the end of the tank up and extending the plastic liner into a temporary spigot.

age of the windows and doors. Refer to the chart on p. 120 to see what factor you need to multiply the footage by. Add 10 percent to this number for defects and waste, and you've got your siding total.

There are bound to be some defects in wood siding. Some are easy to spot, and some aren't. Some defects can be dealt with, and some can't. Discuss the grade of the lumber you want with your supplier, and agree before you place your order what is going to be acceptable.

Among the obvious defects are loose knots. Cut them out if they aren't too numerous, and use the resulting boards in places where shorter lengths are suitable. If loose knots are unacceptable in the grade you ordered, reject the material. In the premium grades, I think it's also fair to reject boards damaged by forklifts or the banding that holds the bundles together.

Warped or crooked boards often can be tamed with some additional fasteners. Bows should be cut out, and twisted or cupped

TIP

Place your best boards near the entry and where the weather does its worst: the south side and the lower courses of the house.

Drying rack. A pair of tilted 2x4s with 80d nails supports the siding as it dries.

boards should be rejected. Wane, the barky surface of the tree that sometimes shows up on the edges of boards, is always attached to sapwood. Depending on the grade you ordered, reject boards with wane.

Checks are cracks in the ends or in the surface of the boards. Surface checking is caused by too-rapid drying and leads to stresses within the lumber: best to reject. End checking is common, but more than a few inches of check indicates too-rapid drying. Accept conditionally.

Fuzzy lumber probably was surfaced with too high a moisture content. If the rough surface is to be the exposed face, go ahead and use the board. If the smooth face is to be exposed, you can sand or plane the board smooth once it's dry. Boards with white specks or decay have been affected by fungus. Paint or stain will hide the specks, but you should cut out decayed portions. If the lumber won't be subjected to wet conditions, further decay won't occur.

Another problem that is virtually impossible to see is case hardening. This condition results from too-rapid drying and creates complex stresses within the lumber. Two signs of case hardening are binding of the sawblade and explosive splitting of the wood when you cut it. Ask for replacement pieces.

Prefinishing with a Dip Tank

Both the California Redwood Association and the Western Red Cedar Lumber Association say you should prime, or seal, the back of your siding (called back priming) and finish it with a water-repellent mildewcide for a natural finish. If you put finish on just the exterior of a piece of siding, it can cup if the backside is exposed to moisture. So for a job like the one pictured here, I dip the boards in Penofin®, a linseed-oil-based clear finish. Penofin provides UV and water protection, and it offers a mildewcide treatment. To speed the process, I submerge my siding in a dip tank. For cedar siding to be painted, I prime it with a stain-blocking oil or latex primer. I prime other species of siding with acrylic latex primer.

My tank's bottom is a wooden I-joist (see the left photo on p. 117). The sides are ½-in. plywood screwed to the I-joist and to 1x2 stiffeners at the top. I made the ends out of 2x10 scraps. One end pivots for draining the tank (see the right photo on p. 117). I fastened a couple of drip brackets to the inside of the upper edge of the tank. After dipping a piece of material for 15 seconds to 20 seconds, I set it into these brackets so that the excess Penofin can drip back into the tank. While this piece drips, I fetch another, submerge it and leave it in the trough while I move the previous piece to the drying rack. At first I lined my tank with 6-mil poly, but it wasn't up to the attacks of ragged grain or the ever-present staples in the siding. So I switched to a cross-laminated polyethylene liner called Cross Tuff®. Cross-laminated polyethylene is amazingly tough and tear resistant. It's typically used under slabs and in walls as a vapor barrier.

My drying racks consist of a pair of cheapo sawhorse brackets, four legs and two predrilled 2x4s with gutter spikes in the holes (see the photo on the facing page). Placing the dripped-dry piece of siding upside down in the rack will prevent beading of the finish on the bottom, exposed edge. This step is really only a concern if you're using a heavy-bodied stain.

As I dip the boards, I study each one for defects. When I encounter boards bad enough to be returned, I set them aside without dipping. After the boards have dried on the rack, I sort them into completely good boards, boards with some defects, and boards that must be cut into short lengths.

The best boards go near the entry and where the weather does its worst: the south side and the lower courses of the house. OK boards go into less visible locations, where they are protected from the weather. The worst stuff goes way up high, out of sight, and out of the weather.

Nailing Recommendation

Nailing patterns vary depending on the profile of the siding. But no matter what the pattern, one thing remains the same: Don't nail through overlapping pieces. To do so will eventually split the siding as it seasonally expands and contracts. Nails should penetrate at least 1½ in. into studs or blocking (1¼ in. for ring-shank or spiral-shank nails). Spacing should be no more than 24 in. o.c. Use box or siding nails for face nailing and casing nails for blind-nailing.

Estimating Coverage

To calculate the amount of material required to side a house, first figure the square footage of the walls minus any openings. Add 10 percent for trim and waste. Now multiply your answer by the appropriate board-foot factor or linear-foot factor to tally the amount.

Felt and Foam

I use 15-lb. felt building paper to protect the sheathing from moisture that might get by the siding. I don't use housewraps, such as Tyvek®, because I'm not convinced they're worth the extra money. What's more, 15-lb. felt makes a great background for the chalklines that I use to lay out the courses of the siding. I generally apply the felt as I go up the building. This technique saves me from ascending the wall twice, and I'm fairly certain I'll have the felt covered before the wind takes it off, or before rain or sun buckles it.

BEVEL

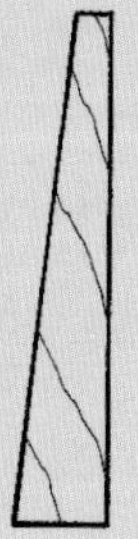

	6 in. and narrower	8 in. and wider	Normal width	Dressed width	Exposed face	Factor for linear feet	Normal width
	1-in. overlap. One nail per bearing, just above the 1-in. overlap.	1-in. overlap. One nail per bearing, just above the 1-in. overlap.	4	3½	2½	4.8	1.6
			6	5½	4½	2.67	1.33
			8	7¼	6¼	1.92	1.28
			10	9¼	8¼	1.45	1.21

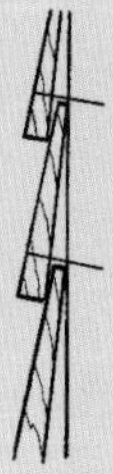

SHIPLAP (Dolly Varden)

	6 in. and narrower	8 in. and wider	Normal width	Dressed width	Exposed face	Factor for linear feet	Normal width
	One nail per bearing, 1 in. up from bottom edge.	One nail per bearing, 1 in. up from bottom edge.	4	3½	3	4	1.33
			6	5½	5	2.4	1.2
			8	7¼	6¾	1.78	1.19
			10	9¼	8¾	1.37	1.14
			12	11¼	10¾	1.12	1.12

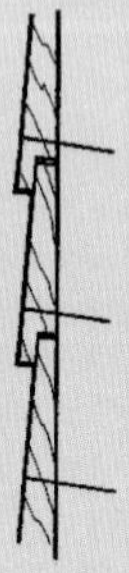

CHANNEL RUSTIC

	6 in. and narrower	8 in. and wider	Normal width	Dressed width	Exposed face	Factor for linear feet	Normal width
	One nail per bearing, 1 in. up from bottom edge.	Use two siding or box nails, 3 in. to 4 in. per bearing.	4	3⅜	3⅛	3.48	1.28
			6	5⅜	5⅛	2.34	1.17
			8	7⅛	6⅞	1.75	1.16
			10	9⅛	8⅞	1.35	1.13

DROP

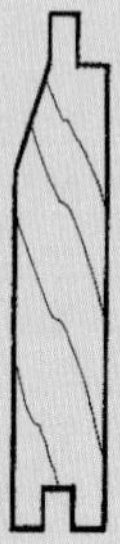

	6 in. and narrower	8 in. and wider	Normal width	Dressed width	Exposed face	Factor for linear feet	Normal width
	T&G pattern — Shiplap pattern. Bind nail T&G patterns; lace nail shiplap patterns, 1 in. up from botton edge.	1-in. overlap. One nail per bearing, just above the 1-in. overlap.	4	3½	2½	4.8	1.6
			6	5½	4½	2.67	1.33
			8	7¼	6¼	1.92	1.28
			10	9¼	8¼	1.45	1.21

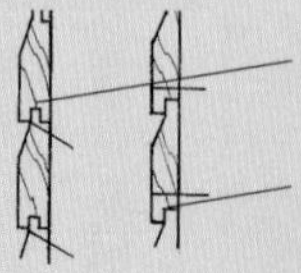

If you're going to apply wood siding over rigid-foam sheathing, remember that foam sheathing has no nail-holding ability. Use nails that are long enough to penetrate through the foam and 1½ in. into the studs.

Two Poles Tell the Layout Story

I use a layout pole and a story pole to lay out the courses of siding before I cut a single board. The layout pole has saw kerfs along its length that mark the distance between each course of siding (see the photo at right). I find this distance by measuring a dozen boards to determine their average width. The industry standard is to leave a ⅛-in. gap between siding boards. But here in southwestern Colorado, it's a rare board that doesn't shrink. That being the case, I leave a ¹⁄₁₆-in. gap between pieces of siding. Once the boards shrink, I'll end up with the recommended spacing. So the marks on my layout pole represent the average width of a piece of siding plus ¹⁄₁₆ in.

The story pole shows where window and door trims will be and where the siding starts and stops. To determine the best layout for the siding, I place the layout pole alongside the story pole, and I slide it up or down to see where the siding will break on window and door trims. The goal here is to avoid narrow rips of siding above or below a window or door. I rarely discover a sublime layout, but I usually identify the difficult areas before climbing a ladder with siding in hand. Sometimes I'll discover that the best way to avoid a nasty string of notched siding boards is to start the first course with a row of siding that has been ripped down to a narrower dimension (see the photo on p. 116). Another strategy that can help avoid a string of complicated notches is to use a wide trim board above or below the windows.

Next I go around the house and jot down the rough measurements between the corners of the building and the openings, such as doors and windows. This list helps me determine how best to use the short lengths of siding that accumulate as the job progresses.

Once I've decided on the elevation of the siding on the wall, I use my layout pole to guide the marks for my chalklines. I use a water level to transfer the marks from one side of the wall to the opposite, and then around the corner and down the next wall.

Layout pole. The author uses a water level to locate the bottom courses of siding on all the walls. Then he marks the subsequent courses with the help of a 12-ft. stick with saw kerfs that correspond to the tops of the siding boards.

Corner Details

Although the variations are endless, there are two basic types of corner boards for houses with horizontal wood siding: butted corner boards and applied corner boards. The applied variety is the easiest to install; however, it can create voids that invite insects, spiders, and wind-borne grit.

BUTTED OUTSIDE-CORNER BOARDS

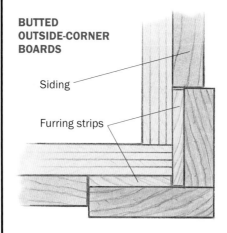

Siding

Furring strips

APPLIED OUTSIDE-CORNER BOARDS

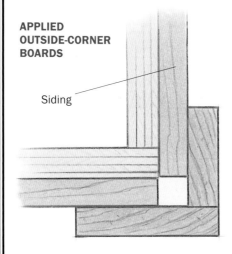

Siding

BUTTED INSIDE-CORNER BOARDS

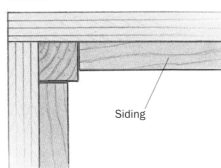

Siding

You've Got Options at the Corners

At outside corners, I prefer to butt the siding to corner boards that stand a bit proud of the siding (see the drawing at left). I typically use rips of siding to make the corner boards, and I fur them out with strips of plywood or rips from framing-lumber scraps.

Applied corners are similar to butted corners except that the installer can be fairly rough with end cuts. Because the corner is applied over these cuts, the ends of the boards will never be seen. This trim detail is much faster, but it has its drawbacks. Where the corner boards overlap certain kinds of siding, such as channel-rustic or beveled siding, you can create hundreds of little nooks for wasps, bees, spiders, and wind-borne grit. And it isn't easy to get paint or stain in there, either. If you choose this corner detail, be sure to seal the bottom of the corner so that it doesn't become a vertical gallery for mice or insects. Incidentally, you've got the same option at doors and windows as you do at corners. You can put casings around windows and doors (my preference) and then run the siding to them, or you can apply casings over the siding.

Mitered corners are another option, but my advice is to avoid them. They are supremely time-consuming, and even when done right, they will eventually open up as the weather does its seasonal work on the wood.

At inside corners I butt the siding to a corner board that is square in section. Applied inside-corner boards are faster, but they don't look as good and invite the same problems as applied outside-corner boards.

Fasteners

Within budget constraints, use the best fasteners available, especially in wet climates. My fasteners of choice are screws. I've used both square-drive stainless-steel and Phillips-drive brass flathead screws to affix siding to houses over ½-in. plywood sheathing. When the sheathing is this thick, I don't bother to break the siding over a stud because the screw threads get such a good bite into the plywood. I predrill all holes in the siding, which eliminates splitting, and I'm delighted in the resulting totally random siding pattern. I don't believe it takes much longer to screw siding in place than to nail it by hand. It probably costs no more if you consider the savings in material by not trimming the siding to break on a stud. The better lumber yield compensates for the extra time. What's more, using screws makes it a lot easier to replace damaged siding or to open a wall for remodeling.

Siding nails, in my order of preference, are stainless steel, hot-dipped galvanized, and aluminum. Spiral or ring-shank nails hold best, and you can generally count on dropping one size in nail length when you use them. The chart on p. 120 shows the recommended nail sizes and patterns for different types of siding.

In my experience, pneumatic nailers can too easily overdrive siding nails. The driver promotes splitting, especially at board ends, which gives water an easy access point. This point said, I must confess to using a ½-in. pneumatic crown stapler with 2-in. stainless-steel staples on my most recent rustic-siding jobs. The siding is ⅞ in. thick, and I ran staples into all the studs and the midpoints between them. The fasteners go in with the speed that makes pneumatic nailers so valuable, and they are virtually invisible on the fuzzy surfaces of rough-sawn cedar siding.

Before committing to using staples, I made several test installations and tried to remove the pieces of siding. The fasteners hung on for dear life, and the boards broke before the staples pulled through the siding or out of the studs. Be aware, however, that staples do not receive official sanction from the trade associations that represent redwood- and cedar-siding producers.

A Three-Man Crew Is Just Right

Putting up siding is almost like painting siding: All the prep work seems to take forever, then the job rolls along at a good clip. Three people on a crew is optimum. One person can cut the siding to length while the other two take measurements, snap chalklines, and fasten the siding. A two-person crew is also efficient, and except for the huge amount of climbing up and down, one person can get it done. A carpenter working solo can support the far end of a piece of

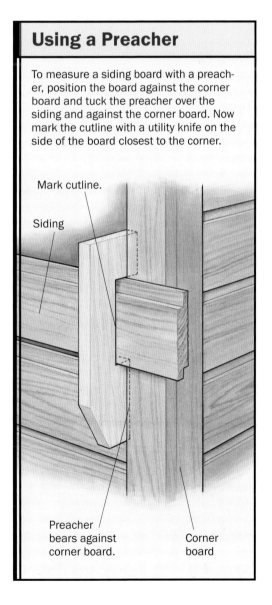

Using a Preacher

To measure a siding board with a preacher, position the board against the corner board and tuck the preacher over the siding and against the corner board. Now mark the cutline with a utility knife on the side of the board closest to the corner.

Mark cutline.

Siding

Preacher bears against corner board.

Corner board

An accurate, portable cutoff rig. A sliding-compound saw, such as the Hitachi C8FB, can make precise crosscuts in siding up to 12 in. wide. Long extension tables support the stock.

Shooting board for acute angles. A straightedge screwed to a pair of fences guides a circular saw through repetitive cuts that are beyond the swing of the sliding compound saw.

siding with a J-shaped hanger made of baling wire or use a coat hanger.

I use a Hitachi® C8FB sliding saw for making most of my cuts on a siding job. I mount the saw in the middle of a long 2x12 that I place across a couple of sawhorses.

Extension tables and fences flank the saw, supporting long pieces of siding during cuts (see the left photo).

For acute-angle cuts that are beyond the swing of the Hitachi, I use a shooting board to guide my circular saw (see the bottom photo). I screw my shooting board to a pair of fixed fences on a temporary bench top. The fences are 1/16 in. thicker than the siding, which allows me to slip each piece of siding under the fixture to make my rake cuts.

Start installing the siding with the bottom board, ripped to whatever width you determined by way of the layout stick and the story pole. The top edge of the board should align with your chalkline. I like the pieces of siding to abut the corner boards snugly, but I don't force them into place. Forcing the boards will push the corner boards out of alignment.

A piece of wood with a notch cut in it, sometimes called a preacher, is a superb little tool for marking exact lengths (see the drawing on p. 123). But using a preacher takes more time than simply measuring because you must lift the siding into place, mark it (use a utility knife), take it back down to the saw bench, and then put it back up. A remodel job that has out-of-plumb corner boards is the perfect place to use the preacher because the cuts won't be quite square.

As I install the siding, I use a urethane caulk to seal the butt joints wherever the siding boards abut a window casing, a door casing, or a corner board. I also use the urethane caulk on the back of the siding to secure any loose knots. The caulk will glue the knots in place.

If I'm using nails or staples to attach the siding, I mark the stud layout on the tar paper with chalklines. Then I break the siding over a stud at butt joints. At each joint, I put

Keep water out at the joints. The author uses narrow strips of waterproof membrane to prevent water from working its way behind the butt joints in the siding.

Weatherize the penetrations. The holes made by hose bibs, pipes, and junction boxes need to be sealed just like doors and windows. This outlet box has a metal flashing folded over its top, plus a flap of waterproof membrane to seal the joint at the wall. A bead of urethane caulk seals the sides and bottom.

a narrow strip of self-adhering waterproof membrane behind the joint (see top photo). The bottom end of this material is led out and over the top edge of the previous piece of siding, preventing water from going through the joint and behind the siding. Tar paper is often used for this detail, but I prefer waterproof membrane because it's less likely to become brittle and crack at the crease. I trim the little flap that protrudes below the siding with a utility knife. Inevitably, there are protrusions such as hose bibs, light fixtures, and outlet boxes that penetrate the siding. These situations are good opportunities for trim carpenters to show

how fastidious their scribe cuts can be. We mark the cut with the siding held in place, and we use a jigsaw to notch the siding. A bead of urethane caulk seals the edges (see the bottom photos above).

I don't like to see a row of butt joints on a wall, so I make sure to stagger them by at least three rows on the same stud. Try to avoid butt joints near an entryway, and never butt trim boards over an entry. This advice would seem obvious, but I see the wrong technique over and over again. Do the splice somewhere else if the trim can't be done with one piece, and if a splice is needed, do it with a scarf joint.

Sources

Louisiana-Pacific
800-648-6893
www.lp.corp.com

Manufactured Plastics
719-487-7373
www.mpdplastics.com

Cross-Tuff Performance Coatings
707-462-7333
Penofin

Trouble Spots Need Flashing

Unless they are protected by roof overhangs, the head casings atop windows and doors should be sealed against the weather with a flashing. I prefer to use copper flashings, but they're pricey. If copper flashings are beyond the budget, galvanized sheet metal, though bright and ugly, is a lot less expensive. These days, there is quite a selection of colors in baked-enamel sheet metal. If the job budget won't buy copper, it will usually accommodate the baked enamel, such as the ones we used on this job.

Where a roof meets a sidewall, you have several options. First, let the roofer roof, then side over his step flashings. Run the building paper over the flashings, but don't put any nails through them. If you do nail the flashings, you make it impossible for a roofer to jockey the flashings around at re-roof time.

Your second option is to put up the siding before the roofing, making certain your fasteners are 2 in. to 3 in. up from the bottom of the siding and not so tight that the roofer can't slip his flashing under the siding. The third choice is to fasten a strip of counterflashing in place, side over it, and let the roofer slip his flashing under yours.

Siding often suffers at this intersection, usually because it has been applied too near the roof. Capillary action can draw water up between the siding and the wall. This process can stain the siding, and it can promote rot. Leave a minimum of 2 in. between the siding and the roof. More if possible, say 4 in. to 5 in., especially if the roof is exposed to heavy weather. The greater clearance is insurance against saturation from mounded snow, soggy leaves, and the buildup from additional roofs.

Divert runoff away from sidewalls. A gable roof that abuts a sidewall can direct water behind the siding at the junction along the roof's edge. Thwart the problem with a flap of flashing that directs runoff away from the wall.

If your sidewall continues beyond the roof's eave, fasten a piece of flashing along the roof's edge to keep water from going behind the siding (see the photo above). Confer with your roofer before proceeding with any of these options because he's the one who will have to stand behind the roof. And make sure that both you and the roofer are using the same kinds of flashing materials. For example, don't mix copper and galvanized.

Felix Marti designs and builds home with an emphasis on energy efficiency and low maintenance. He lives in Ridgway, Colorado.

Installing Wood Clapboards

■ BY RICK ARNOLD AND MIKE GUERTIN

We've installed clapboards just about every way but upside down. We've tried different nailing locations and patterns and used various hand-driven and pneumatic fasteners. We've installed clapboards from the bottom up and from the top down, and we've organized our installation teams several ways. Although we wouldn't call any of our experiments failures, we have found a layout method, fastener and fastening pattern, crew size, and details that ensure good performance and efficient installation.

A Few Details before Starting

Before we lay out and install clapboards, several details need to be in place. We have up whatever housewrap we're using, and install the corner boards (see drawings on facing page). Corner boards provide a stop for the clapboards so that they don't have to be mitered or coped. We make our inside corners out of 1x2 or square 5/4 stock. Anything less would be too thin and leave the painter no room to caulk between the inside corner and the ends of the clapboards.

Outside corner boards avoid the need for delicate mitered joints at locations easily damaged by lawnmowers and other equipment. We make ours from 1-in. stock and prime any raw edges before installing the siding. Doing so minimizes shrinking and swelling and just about eliminates stains from dissolved tannins leaking to the face.

For a heavier look, we sometimes build outside corners from 1x3 furring and apply 1x6 finish corner boards over them after the siding goes on. We nail the outside edge of the finish corner board to the clapboards every 16 in. to 24 in., being sure to nail through the thick bottom of the clapboards. The clapboards support the overhanging corner board, and nailing is less likely to crack it. Overlapping clapboards with the corner board in this way is a more watertight detail, but the nooks where the clapboards go under the corners can provide homes for insects.

When we apply the corner boards, we let them run long. Then we trim them to length once the water table—the horizontal board where the wall meets the foundation—has been installed (see the sidebar on p. 132).

On gable ends without overhangs, we install 1x6 rake boards over 1x3 furring that runs flush with the top of the roof sheathing. We leave enough room to slide the clapboards at least 3 in. under the 1x6 (see the bottom drawing on facing page). On overhanging rakes, we cut the clapboards to the rake angle and butt them to the soffit. Then we trim this joint with a piece of 1x3.

Careful Layout Is Crucial to a Good Job

The common clapboard in our area is $\frac{3}{8}$ in. thick at the butt end and $5\frac{1}{4}$ in. wide. Clapboards typically have 4 in. of exposure to the weather, but we tweak the layout so that the bottom of one course lands directly on top of the windows and the bottom of another course matches the bottoms of the windows (see the photo on p. 135). We also want no more than a $\frac{1}{8}$-in. variation between adjacent courses. With luck, we can make the bottom of a course fall directly on top of the doors, too. Good layout results in

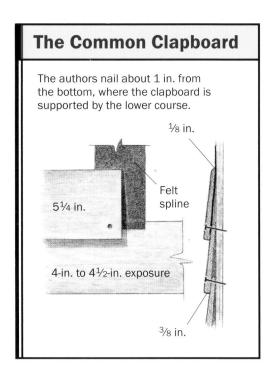

The Common Clapboard

The authors nail about 1 in. from the bottom, where the clapboard is supported by the lower course.

$\frac{1}{8}$ in.

Felt spline

$5\frac{1}{4}$ in.

4-in. to $4\frac{1}{2}$-in. exposure

$\frac{3}{8}$ in.

Outside Corners Serve Two Purposes

They are a way to turn a corner without using a compound miter, and they protect the clapboards from damage at this exposed location. Overlapping 1x4s form the simplest corner. For a heavier look, nail on a subcorner of 1x3 furring, butt the clapboards to the 1x3, and apply a finish corner made of 1x6 that laps the clapboards and the water table.

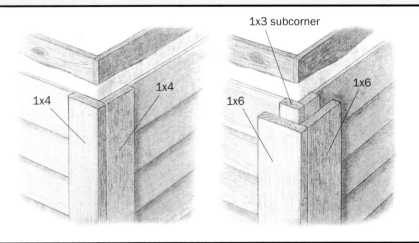

1x3 subcorner

1x4

1x4

1x6

1x6

Inside Corner Boards Make Coping or Mitering the Clapboards Unnecessary

Butt the clapboards to a vertical piece of 5/4 square stock or 1x2 at the inside corners. Both options leave plenty of room for a neat caulking job.

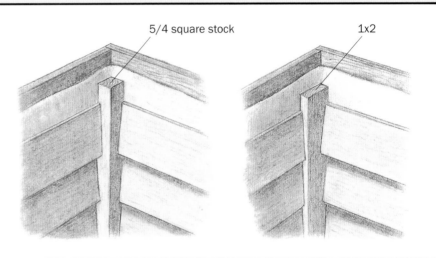

5/4 square stock

1x2

A Neat Finish Where Clapboards Meet the Gable

On walls with no overhanging gable, nail 1x3 furring to the house and go over this furring with a 1x6 finish rake board. The tops of both are flush with the top of the roof sheathing. The roof shingles extend over these boards, and the clapboards tuck into the space behind the 1x6. On houses with an overhang, cut the clapboards to the rake angle, butt them to the soffit, and cover this joint with a trim piece.

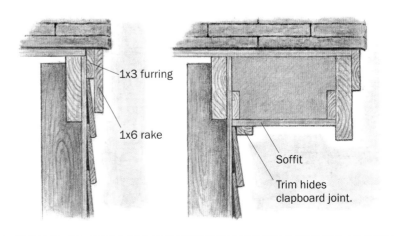

1x3 furring

1x6 rake

Soffit

Trim hides clapboard joint.

Layout with a Story Pole

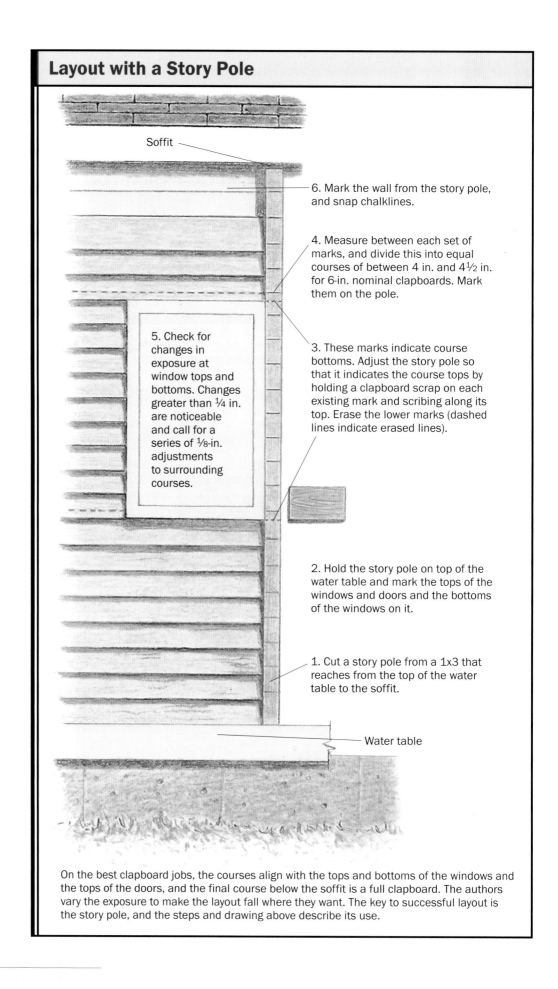

Soffit

6. Mark the wall from the story pole, and snap chalklines.

4. Measure between each set of marks, and divide this into equal courses of between 4 in. and 4½ in. for 6-in. nominal clapboards. Mark them on the pole.

5. Check for changes in exposure at window tops and bottoms. Changes greater than ¼ in. are noticeable and call for a series of ⅛-in. adjustments to surrounding courses.

3. These marks indicate course bottoms. Adjust the story pole so that it indicates the course tops by holding a clapboard scrap on each existing mark and scribing along its top. Erase the lower marks (dashed lines indicate erased lines).

2. Hold the story pole on top of the water table and mark the tops of the windows and doors and the bottoms of the windows on it.

1. Cut a story pole from a 1x3 that reaches from the top of the water table to the soffit.

Water table

On the best clapboard jobs, the courses align with the tops and bottoms of the windows and the tops of the doors, and the final course below the soffit is a full clapboard. The authors vary the exposure to make the layout fall where they want. The key to successful layout is the story pole, and the steps and drawing above describe its use.

a look that sets clapboards apart from vinyl siding, whose layout shows no regard for window and door height.

By the way, layout is quick and easy if you set all the windows at the same height. If their height varies, plan the differences between windows to match clapboard-course increments.

Make sure the top of the water table, usually 1x6 stock, falls on a 4-in. to 4½-in. increment from the bottom of the windows and that its bottom overhangs the foundation by 1 in. If the foundation steps down as the grade changes, we lower the water table by course increments.

The story pole is the crucial layout tool. We cut it from 1x3 furring and make it long enough to reach from the top of the water table to just beneath the soffit. On this job, we needed two story poles because a second-floor overhang made it impossible to use one that was two stories high.

With the story-pole bottom atop the water table, we mark the tops and bottoms of each window and the tops of the doors on the pole. We indicate which mark belongs to which window or door to avoid confusion. These window and door locations are the primary marks.

The tops of the clapboards have to be marked on the wall. If the bottoms were marked, each clapboard would cover up the mark for the next one. The primary marks indicate clapboard bottoms. To make the adjustment, we hold a piece of clapboard to each mark and to the bottom of the pole and draw a line at its top. We then erase or cross out the original marks. Now we have primary marks indicating the tops of clapboard courses that match windows and doors.

Starting at the bottom of the story pole, we measure to the next higher mark and divide this distance into equal courses of 4 in. to 4½ in. We do the same thing between each pair of primary marks. Sometimes the space between primary marks is too awkward to divide, say 10 in. There's no way to match a 4-in. course to that space. Usually, it's on the back or side of the house with a shorter bathroom or kitchen window or it's at the top of a door, so it doesn't concern us. We erase the out-of-sync primary mark to avoid confusion, knowing that we'll be notching a clapboard to fit around a window that is off-layout. After we divide the sections, we check for abrupt changes in exposure; those greater than ¼ in. are noticeable. The biggest differences occur just after a primary mark.

Because a primary mark indicates a door or window, we can't change it. We blend the change over a couple of the surrounding courses, making it less obvious. For example, say the courses change from 4 in. to 4⅜ in. We would make one of the 4⅜-in. courses 4⅛ in., then a couple courses at 4¼ in. and a couple at 4⅜ in. To make up the difference, we mark a few at 4½ in. and go back to 4⅜ in. Once we select the final course marks, we square them off and erase any errant marks.

With the pole complete, we use it to mark the sides of all corner boards, windows, and doors. We mark and snap all the lines within reach and begin installing clapboards. As we raise the staging, we mark and snap the rest.

One carpenter nailed on corners and the water table, another cut the starter clapboards. He planned the cuts so that the joints are randomly staggered. Efficient crew members leapfrog each other, cutting or measuring ahead so that there is no downtime.

The Water Table Is the Starting Point

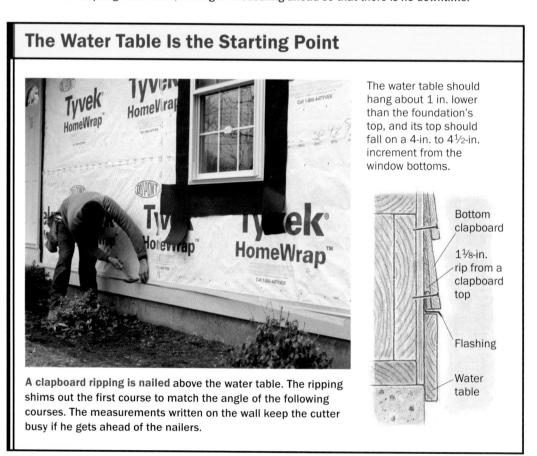

A clapboard ripping is nailed above the water table. The ripping shims out the first course to match the angle of the following courses. The measurements written on the wall keep the cutter busy if he gets ahead of the nailers.

The water table should hang about 1 in. lower than the foundation's top, and its top should fall on a 4-in. to 4½-in. increment from the window bottoms.

Bottom clapboard

1⅛-in. rip from a clapboard top

Flashing

Water table

Three Carpenters Make an Efficient Crew

One carpenter cutting and two nailing keep the momentum going the best. We first install the water table and cap it with an aluminum drip cap that extends about 1½ in. up the wall and hangs over the water table by ¼ in. or so. Then we measure short spans between windows and count how many courses high the windows are. We write these numbers in crayon on the wall. The cutter can spend his downtime pre-cutting those sections. The cutter begins by squaring the factory ends and setting up starter packs of staggered lengths for long sections of wall. We insist on random staggers. Nothing's worse than seeing equal 1-ft. staggers forming a pattern that catches your eye.

We nail a 1⅛-in. ripping from the top of a clapboard to the wall above the water table (see sidebar on facing page). This shims the bottom of the first clapboard out to match the succeeding courses. The measurement 1⅛ in. is the common overlap of 4-in. ex-posures of 5¼ in. (finished size) clapboard, minus the saw kerf. We use the leftover 4-in. bottom rip as the last piece under the soffit.

The nailers tack up the starter clapboards and measure the subsequent pieces before nailing off. The cutter can ready pieces with no downtime. Installers often tack the clapboards on a whole face and finish nail when they crank the staging down. By keeping ahead of each other, by precutting or by premeasuring, the cutter and nailers always have something to do.

All raw wood is primed. Water entering through end grain gets under the surface and causes paint to peel. Back-priming and end-sealing minimize this problem.

Even when we've sanded, we've had paint fail on smooth clapboards.

Details for Appearance and Weather Resistance

There are several reliable details we use on every job. Although many siders bevel-cut their joints, we butt ours. We have no shrinkage problems with the dry finger-joint primed material we use, and square cuts look better over time. We keep a pail of primer handy and dab cut ends to seal out water. We insert tar-paper or aluminum step-flashing cards behind each seam. They lap the lower clapboard by ½ in. or so and shed any water that penetrates the joint. (see the top photo on p. 134). We keep the clapboards

Tar-paper splines ensure a leakproof joint. Placed under each butt joint and at the windows, the splines lap the lower course by ½ in. and guide any water that enters the joint to the outside of the clapboards.

Tar-paper splines lap each other from top to bottom. Finally, they lap the top of a lower clapboard. Any water that gets behind the siding is shed back to the outside.

¹⁄₁₆ in. to ⅛ in. back from windows. We've had clapboards swell and pinch the windows so that the sash wouldn't move.

A bead of paintable silicon caulk around the window seals out water and lets the clapboards move without distorting the jamb. In high-wind areas where leakage from driven rain is a concern, we caulk the clapboard ends before butting them to the corner boards or casings.

We put tar-paper or bituminous-membrane splines around window flanges for protection from water. The head spline laps the side splines, which lap the bottom spline. We lap the bottom spline ¼ in. to ½ in. over the top of a lower course of clapboard (see the photo at left).

Where a wall, such as a dormer sidewall, meets a roof, we lay a clapboard flat on the shingles with its thin edge against the wall to act as a spacer. We then cut, prime, and nail the clapboards to the sidewall, tight against the spacer. When we remove the spacer, there is a clean ¼-in. space between the siding and the roof shingles. Without this space, the siding could wick up water, blistering paint and hastening decay.

When we get to the top of a window, the course just below butts the casing on two sides. The tops of these clapboards extend 1¼ in. higher than the window, but there is nothing above the window. To shim out the next course, we rip the top of a piece of clapboard to 1⅛ in., just as we did above the water table. We cut it to fit between the projecting tops of the lower course, nail it in place above the window and move to the next course (see photo below).

Clapboards come with one side rough and one side smooth. We guarantee finishes only on rough-side-out installations. Smooth-side-out installations require labor-intensive hand-sanding to break the shiny surface left by the planing process or the primers won't bond. We've found that the smooth side of preprimed clapboards needs a light sanding before finish-painting. Even when we've sanded, we've had paint fail on smooth clapboards.

The top of a clapboard fills the space above the windows. This keeps the next course from bowing inward as it's nailed. With vinyl windows such as these, the authors leave a ⅟₁₆-in. to ⅛-in. expansion gap between the windows and the clapboards and caulk it.

Nailing Wrong Means More Than Hitting Your Thumb

Fastener location and pattern are controversial issues. We'll admit it: We nail in the wrong place. We nail too low; everybody we know nails too low, and they know it, too. You're supposed to nail through the lower part of the clapboard just above the top of the clapboard beneath. In theory, this gives the boards freedom to expand and to contract in height. If you nail too low, as we do, you'll trap the clapboard with nails high and low, and the clapboards will split as they move. In theory.

The reality is that we've never had a problem when we nail too low, say within 1 in. of the bottom. But we have had problems when we nail "properly." The upper board splits at that point because it has no support from below. Sure, we could nail loosely to try to avoid splitting, but what about the customers who want the nails set and the holes filled? Splits galore. We're not the only ones who nail wrong. We've remodeled homes built in just about every decade of the past 250 years and found similar "poor" workmanship in most. Nuff said.

The proper nailing pattern for clapboards is just as controversial. Should you nail randomly, or should you nail into the studs? The only time we purposely nail into the studs is when a house is sheathed with foam board or when the customer thinks it's a quality issue. We sheathe most of our homes with $\frac{7}{16}$-in. oriented strand board (OSB). OSB holds nails well, so we don't have to hit the studs.

Nailing into the studs causes aesthetic defects. First, the uniform vertical lines of nails belie the horizontal nature of clapboard. Second, the clapboards are drawn more tightly to the wall at the nails. Nailing into the studs creates noticeable waves in the siding. And a 2-ft. stud spacing makes the problem worse. The only solution is random nailing. When we argue the issue with other builders, they insist that we're cheating, that random nailing is easier. Actually, it's not. Easy is having marked vertical lines to nail on. It's difficult to be carefully random and to maintain regular 12-in. to 18-in. nail spacing. We take great care not to have one nail atop another. Random nailing works well for us because we locate our fasteners thoughtfully.

We Finally Found the Right Fasteners

We have tried several fasteners over the years; some worked, some didn't. Pneumatic staples driven parallel with the grain leave tiny slots that need to be filled before painting. The galvanized staples hold well, and the gun countersinks them well. But we rarely use this method because we install the clapboards rough side out, and the smooth, puttied slots stand out. On smooth-side-out installations, we might consider staples.

We've tried pneumatically driven galvanized ring-shank box nails. We could adjust the gun so that the nails would either be set or be flush. Nailing was fast. But the electro-galvanized finish wasn't thick enough, and they'd rust after a year.

Hot-dipped galvanized ring-shank shake nails were the old standard; we used them for years. We'd set the nails flush and paint over them. Two problems developed. Hammering the nails wears off the galvanized finish, causing them to rust. The tannins in red cedar and redwood react with the galvanization and cause streaking that goes right through the finish paint or stain.

Our final solution is stainless-steel ring-shank siding nails set flush. They cost five times what galvanized nails do. That sounds like a lot, but it's $10 vs. $50 on a house. That's money well spent. The waffle pattern

Choosing a Wood Species for Clapboards

We've applied clapboards of several wood species over the years. Red cedar is the most common, and redwood a sometime substitute. We've seen pine and spruce on occasion, and mahogany only rarely. Knotty pine has a poor record of splitting, warping, and shrinking. We avoid it completely. Radially sawn clear pine might be better, but we haven't used it. Spruce is a newcomer in our region. Although we haven't tried it yet, we've heard good reviews from those who have. Mahogany is seldom available, and we've found it hard to nail without splitting. Otherwise, it performs nicely without shrinking or warping, and it takes finishes well.

Red cedar comes in several grades. Clear vertical grain performs the best and is the most costly. A and better performs well and is what we use the most. Because most of our customers prefer to paint clapboards, we pay a little extra for factory-preprimed stock. Preprimed wood resists absorbing moisture from the back that can cause the paint to blister and peel. It nearly eliminates cupping and extractive bleed. Before preprimed material was readily available, we back-primed the clapboards ourselves to get a better finished product. We don't miss the mess. Some of our first preprimed clapboard projects still sport their original paint and after nearly 13 years look like new.

To be both environmentally and cost conscious, we began installing preprimed, finger-jointed clapboards 12 years ago and never looked back. We no longer have to use five bundles of 4-footers on the back of the garage; finger-jointed clapboards are manufactured in 16-ft. lengths. Occasionally, there's a bad section, but we can easily cut it out before installation. We've had to replace only two or three boards in the last 150,000 lin. ft.

on the nail head blends in with the rough texture of the siding, so with random nailing, they almost disappear.

Real Clapboards vs. the Vinyl Alternative

In our area, ½-in. by 5¼-in. finger-jointed preprimed A and better clapboards cost $160 per square. Good-quality vinyl siding goes for $60 per square. It takes about 50 percent longer to install clapboards than vinyl siding. Then there's the cost of painting or staining the entire house every 5 years. Even in conservative, slow-to-change New England, we're seeing less clapboard siding going on new homes. Economics and high maintenance are taking their toll.

Still, vinyl siding just doesn't look as good. It fits awkwardly around windows. Seams are obvious when viewed from the wrong direction. There's debris-catching J-channel everywhere. Vinyl creaks from expansion when the sun hits it. The best vinyl siding is said to look almost like clapboards. Almost. But there's nothing like the real thing.

Rick Arnold and Mike Guertin are contributing editors to Fine Homebuilding *magazine and the coauthors of* Precision Framing. *In addition, Rick Arnold is the author of* Working with Concrete *and Mike Guertin is the author of* Asphalt Roofing, *all published by The Taunton Press.*

10 Rules for Finish Carpentry

■ BY WILL BEEMER

My first construction job was as a trim carpenter's helper during school summer vacation. My boss had always worked solo, but as he got on in years (I'm older now than he was then), he wanted help moving his tools and materials. All I did that first summer was fetch and carry; I wasn't allowed to measure, cut, or nail. I was told to observe. In doing so, I learned that finish carpentry is essentially a visual exercise.

Finish carpentry makes the eye work hard and skip over imperfections. At this point, the framing carpenter has made the house plumb, level, and square. Or not. A good framer can ease the finish carpenter's job by providing plumb walls and plenty of blocking for nailers for attaching trim. Or not. But even if the framer couldn't read a level and even if he added no more blocking than was absolutely necessary, the finish carpenter's job is to make the doors, windows, and cabinets work, and to make the house look good.

Finish carpentry is more than interior trim. It includes siding, decking, and even roofing—anything the owner will see after moving in. Rough carpenters evolve into finish carpenters by learning how to measure, mark, and cut more accurately. With practice, splitting the pencil line with a saw-cut and working to closer tolerances become second nature.

Perfect miters are only part of finish carpentry. Finish carpenters must develop an eye for proportion and detail. They must learn to visualize the steps that lead to the finished product. I teach these skills to novice carpenters at the Heartwood School in Massachusetts. To make learning them easier, I've organized the following 10 rules of thumb.

1. Avoid Using Numbers

It is usually more accurate to hold a board in place to mark its length rather than to use a tape and involve numbers. Sometimes, using a ruler or tape is unavoidable. I'll use a tape measure on a long piece that's too difficult to mark in place, but generally I don't like tapes. A tape can flex and change shape, and

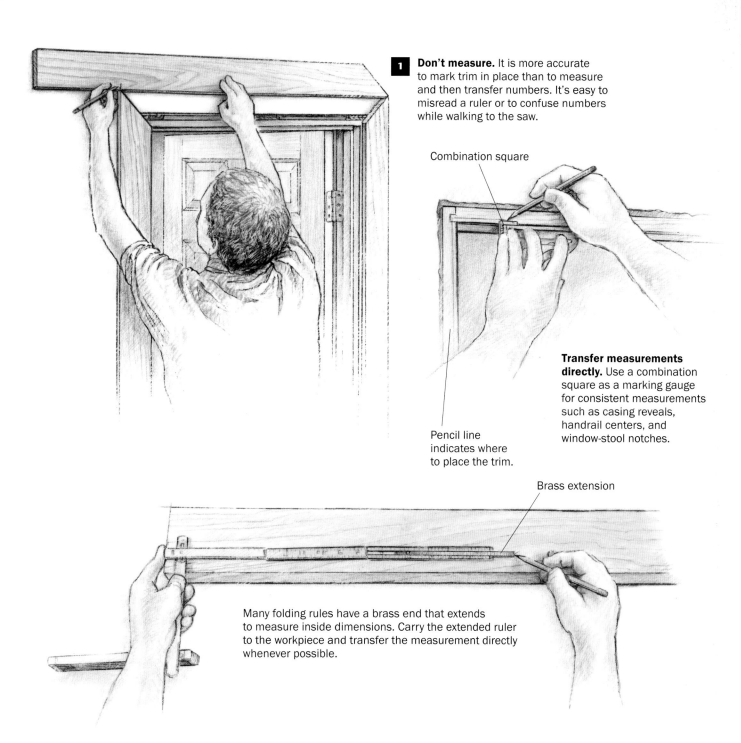

1 **Don't measure.** It is more accurate to mark trim in place than to measure and then transfer numbers. It's easy to misread a ruler or to confuse numbers while walking to the saw.

Combination square

Pencil line indicates where to place the trim.

Transfer measurements directly. Use a combination square as a marking gauge for consistent measurements such as casing reveals, handrail centers, and window-stool notches.

Brass extension

Many folding rules have a brass end that extends to measure inside dimensions. Carry the extended ruler to the workpiece and transfer the measurement directly whenever possible.

the movable end hook bends easily, affecting accuracy.

A rigid rule is better than a tape for measuring lengths less than 6 ft.; hence, the 6-ft. folding wooden rule takes over during trim and cabinet work. The best folding rules come with a sliding brass extension that makes taking inside measurements easy. Open the rule to the greatest length that fits between the points to be measured and slide out the brass extension the rest of the distance. Hold it at that length, and carry it

to the board to be cut. No need for numbers. Just mark the board from the extended ruler. A combination square or a wood block of known dimension is the best way to lay out the small measurements needed for reveals and other spacings. Learn what dimensions are built into the tools you use. A carpenter's pencil is ¼ in. thick; you can use it as a spacer for decking. The pencil lead is 1/16 in. from the edge of the pencil, so it can scribe 1/16-in. increments. The body of a folding rule is 5/8 in. wide. The blade of the standard combination square is 1 in. wide, and its body is ¾ in. thick.

A door or window should be cased without the use of a tape. Lightly mark the reveal on the jamb with a pencil. Square-cut the bottoms of the casing legs, hold them up to the jamb, and mark the top cuts from the reveal lines. Cut the legs and tack them in place. Miter one end of the head, and holding it upside down over its final position, mark the other end to length.

2. Use Reveals, and Avoid Flush Edges

Wood moves—as it dries out, as the house settles, as you cut it, and as you're nailing it up. It's almost impossible to get flush edges to stay that way. That's why, for example, carpenters usually step casing back from the edge of door and window jambs. Stepping trim back to form reveals causes shadowlines and creates different planes that make it harder for the eye to pick up discrepancies. If a casing is installed flush to the inside of

a jamb, it may not stay that way. The eye will easily pick up even a ⅟₁₆-in. variation from top to bottom. If the casing is stepped back ¼ in. or ⅜ in., this variation will not be nearly as evident and will be hidden in shadow much of the time. Separate discrepancies, and they become less evident.

In years past carpenters by necessity used trim materials of different thicknesses; planers were not in widespread use. You rarely see mitered casings in older houses because differences in material thickness are obvious in a mitered joint. Instead, the casing legs butt to the head, which runs over and past the legs by ⅜ in. or so. This way, the carpenter didn't worry about the length of the head casing being exact or the side casings noticeably changing width with changes in humidity. The head casing is usually the thicker piece so that the shadow it casts makes it appear to be a cap. Carpenters in the past often placed rosettes at the upper corners and plinth blocks at the bottoms of door jambs. The casings and baseboards butted to them. The variations in thickness of these boards were lost in the overwhelming presence of the thicker plinths and rosettes.

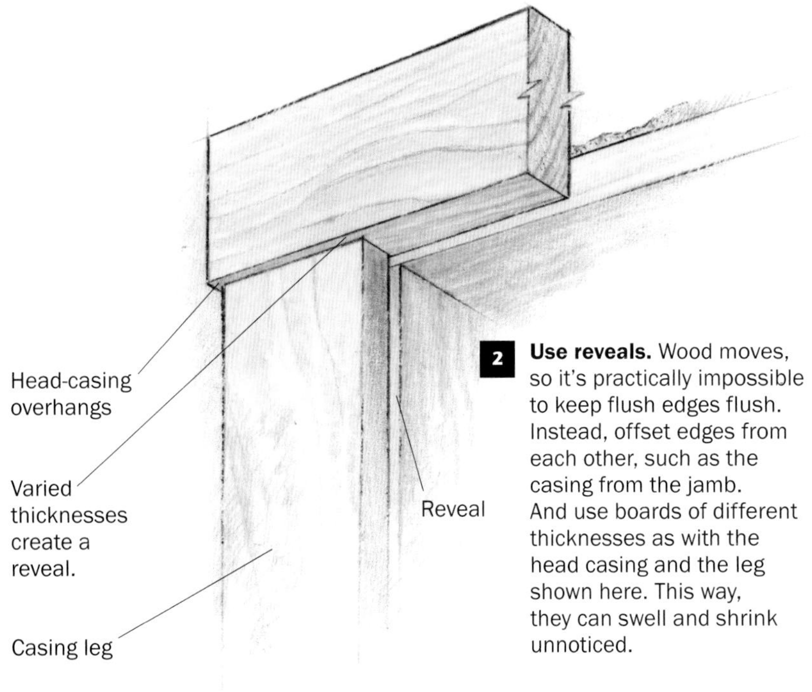

3. Split the Difference

If you're running courses of material between two diverging surfaces, it's obvious that if you start out working parallel to one, you won't be parallel to the other. In the case of decking, roofing, or siding, you can slightly adjust the gap or coverage at each course so that the courses are parallel to the other surface when they reach it.

This adjustment is easily figured. Say that you're shingling an old house, and the roof measures 135 in. from ridge to eave on one end and 138 in. on the other. Divide one of these figures by the ideal exposure per course, 5 in. for normal three-tab shingles. Thus, 135 in. divided by 5 equals 27; this is the number of courses at 5 in. per course.

Head-casing overhangs

Varied thicknesses create a reveal.

Casing leg

Reveal

2 **Use reveals.** Wood moves, so it's practically impossible to keep flush edges flush. Instead, offset edges from each other, such as the casing from the jamb. And use boards of different thicknesses as with the head casing and the leg shown here. This way, they can swell and shrink unnoticed.

At the other end of the building, 138 in. divided by 27 yields 5⅛ in. Lay out each side of the roof using the two different increments, and snap chalklines between them. With these adjustments, the chalklines start out parallel to the eaves and end up parallel to the ridge.

It's difficult for the eye to pick up even a ¼-in. change in exposure from one course to another, especially high up on a house. By accounting for discrepancies early and gradually, your adjustment should rarely have to be greater than that.

In cases where the gap or coverage is not adjustable, as in tongue-and-groove flooring, you have to make up part of the discrepancy at the start and the rest at the end. Say you're installing flooring between two walls that are 1 in. out of parallel, and you're leaving a minimum expansion gap of ½ in. between the flooring and the wall. Make the expansion gap 1 in. at each side of the wide end of the room and ½ in. at each side of the narrow end. Shoe molding and baseboards cover the gap (bottom drawing). If you're using a one-piece thin baseboard, you'll have to rip tapered floorboards at the start and finish to keep the expansion gap narrow and parallel to the wall.

In any case, you'll want to use boards as wide as possible as your starting and ending courses. Measure the room width at both the wide and narrow ends, and subtract the expansion gaps. Divide these measurements by the floorboard width. Multiply the remainders by half the board width. These will be the widths of the starting and ending strips at the wide and narrow ends of the room. If these strips are narrow, try adding half a board width. As long as these sums are less than full board widths, use them for the starting and ending strips. This keeps the converging lines of the baseboards and the first and last seams as far apart as possible.

3 **Diverging lines are obvious** mistakes. With shingles or lapped siding, diverging starting and ending points can be hidden a little at a time by slightly tapering the course widths. But this technique doesn't work with other materials, such as tongue-and-groove flooring, whose course can't be easily varied.

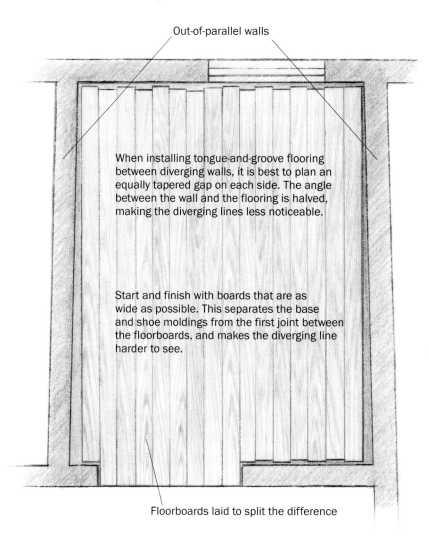

Out-of-parallel walls

When installing tongue-and-groove flooring between diverging walls, it is best to plan an equally tapered gap on each side. The angle between the wall and the flooring is halved, making the diverging lines less noticeable.

Start and finish with boards that are as wide as possible. This separates the base and shoe moldings from the first joint between the floorboards, and makes the diverging line harder to see.

Floorboards laid to split the difference

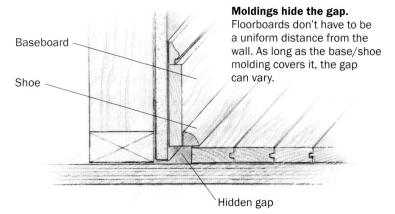

Baseboard

Shoe

Moldings hide the gap. Floorboards don't have to be a uniform distance from the wall. As long as the base/shoe molding covers it, the gap can vary.

Hidden gap

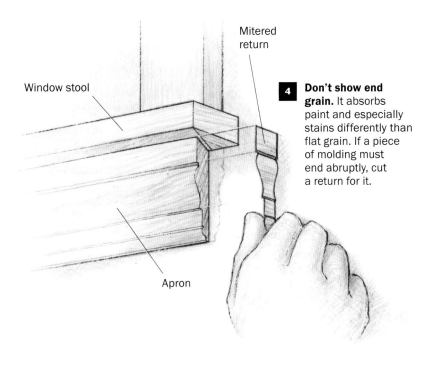

Window stool

Mitered return

Apron

4 **Don't show end grain.** It absorbs paint and especially stains differently than flat grain. If a piece of molding must end abruptly, cut a return for it.

4. Avoid Exposing End Grain

End grain will always absorb stain and paint differently than face or side grain; even if left natural, end grain reflects light differently. Unless you want to emphasize this difference, as with through tenons in furniture, it's best to plan your installation to hide end grain or cut mitered returns to cover it up.

A return is a small piece of trim, often triangular in section, that ends a run of molding. Returns are used in traditional finish work on stair treads, window stools and aprons, butted head casings—anywhere a piece of molding doesn't end in a corner. Small returns are difficult to cut on a power miter saw. The blade often throws the return to some dimly lit, inaccessible corner of the room. I cut them with a small miter box and a backsaw.

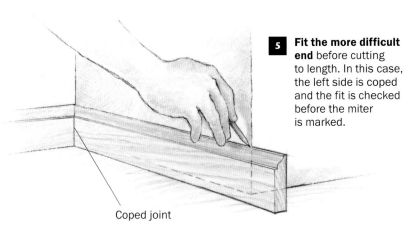

Coped joint

5 **Fit the more difficult end** before cutting to length. In this case, the left side is coped and the fit is checked before the miter is marked.

5. Fit the Joint before Cutting to Length

If you're coping or mitering a joint on a piece of base, chair rail, or crown, make sure that joint fits well before you cut the other end to length. You may need the extra length if you make a mistake and have to recut the cope or miter. If you had cut the piece to length before miscutting the cope or miter, you'd be grumbling on your way back to the lumberyard instead of calmly recutting the piece.

6 **Some joints don't need** to be perfect. Base will hide the ugliness where the drywall meets the floor.

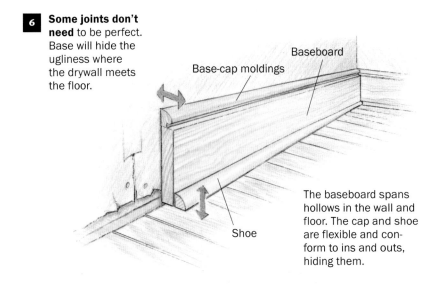

Baseboard

Base-cap moldings

Shoe

The baseboard spans hollows in the wall and floor. The cap and shoe are flexible and conform to ins and outs, hiding them.

6. Don't Be Fussy Where You Don't Have to Be

Learn to think ahead to see if what you're working on will later be covered, which is often the function of moldings. At the intersection of wall and floor, for example, the drywall doesn't have to come all the way down to the floor, nor does the flooring need to meet the wall perfectly, because the baseboard will cover the gap. If the floor or wall undulates, you might be tempted to scribe or fill behind the baseboard to follow the contours. In older houses, however, where walls and floors always undulated, you often see three-piece baseboards, with the thin base-cap molding attached to the wall and following its contour while the shoe does the same on the floor. The thicker baseboard installs quickly and easily because it doesn't have to conform; that's what the shoe and cap do.

7. Plan Your Sequence to Avoid Perfect Cuts at Both Ends

There is usually a sequence of installing trim that requires the fewest perfect cuts. For example, with my method of casing doors and windows, only the last cut on the head need be perfect. Cut this end slightly long, and shave it with the chopsaw until it fits just right. One neat trick here: Push the casing

7 **Trimming a room with baseboard and a minimum of perfect cuts.** By following the numerical sequence in the drawing below, only pieces 2 and 3 require perfect cuts on both ends. The chance of error is reduced by first coping them and then holding them in place to mark their lengths. The copes are planned so that any cracks will be less obvious to people entering the room.

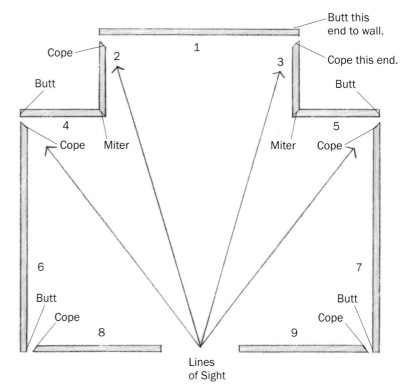

Coped joints. The first piece is butted to the wall. The second piece is mitered as for an inside corner, but the mitered end is cut off where it meets the molding face, leaving a negative of the profile that fits perfectly over the butted piece.

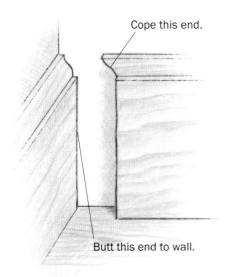
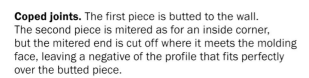

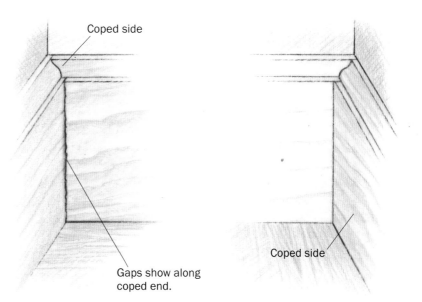

Coped joints look different from different angles. If a coped joint opens up, the crack will be obvious when viewed parallel to the uncoped piece and nearly invisible viewed parallel to the coped piece. Plan the coping sequence so that cracks will be less obvious along likely lines of sight. Cracks will also be less obvious if the uncoped piece is stained or painted before installation; raw wood sticks out.

up to the lowered, idle chopsaw blade. Raise the blade without moving the casing, and then make the cut. The teeth are set slightly wider than the body of the blade, so the cut will take off $1/32$ in. If you had installed the head first, you then would have had to make an exact miter cut on each casing leg to make the joint turn out right.

As with trimming doors and windows, sequence of installation is also important when running trim around a room, whether it's baseboard, chair rail, or crown molding. For instance, I'm right-handed and can generally do a faster, neater job of coping the right end of a board than the left end (coping is an alternative to mitering joints at inside corners; see the bottom drawings on p. 143). Consequently, I often prefer to work from right to left around a room.

As I work my way around a room, especially when running crown molding, I'll often end up with a piece that needs to be coped on both ends, a challenge for even the best carpenters. I try to plan my installation so that this last piece of trim is in the least conspicuous place. If a coped joint isn't perfect, or if it opens up over time, the crack is most visible when viewed at right angles to the coped piece. Wherever possible, I orient the coped pieces so that people entering or using the room won't have right-angle views of them.

8. Parallel Is More Important Than Level or Square

Some rules of carpentry change from framing to finish work. Instead of keeping track of plumb, level, and square, you now must keep finish materials parallel to the walls and floors. The eye sees diverging lines more readily than it sees plumb and level.

The only exceptions are cabinets and doors, which must hang plumb to work properly. If the floor isn't level, trim the door bottoms parallel to the floor rather than leave them level with a tapered gap.

If a deck is out of square, run the decking parallel to the house wall rather than squaring it up to the deck framing. If for some reason two lines must diverge, separate them as widely as possible so that the difference is harder to see.

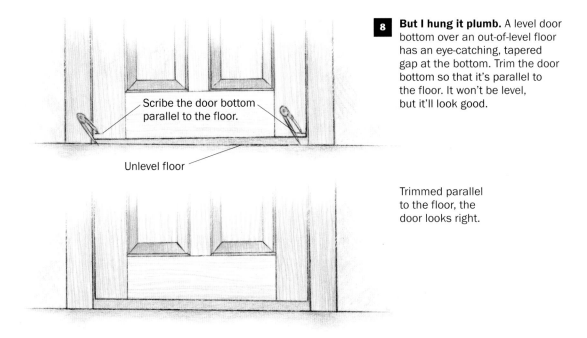

Scribe the door bottom parallel to the floor.

Unlevel floor

8 **But I hung it plumb.** A level door bottom over an out-of-level floor has an eye-catching, tapered gap at the bottom. Trim the door bottom so that it's parallel to the floor. It won't be level, but it'll look good.

Trimmed parallel to the floor, the door looks right.

9 **Nothing is random.** Even something as simple as decking benefits from thoughtful layout. The randomness (left) looks sloppy compared with careful layout (right).

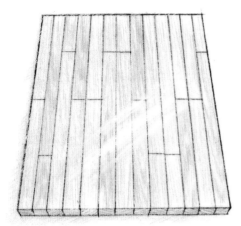

9. Nothing Is Random

Whenever I find myself saying, "It doesn't matter," the red flag goes up. Which end of the board you cut first, which face is out, where you put the nails—all this matters, and the care you put into the details shows up as craftsmanship in the entire job. "God lives in the details," said Mies van der Rohe, and this is especially true in finish carpentry. Occasionally, it won't matter, but you should first consider whether it does. Taking this to the extreme may result in a phenomenon I call "analysis paralysis." As your experience increases and your eye becomes more efficient, it will become second nature to line up your nails in an attractive pattern and to look critically at each board as you carry it to the saw.

10. Finish the Job

A contractor usually has to complete a punch list before final payment is issued, but sometimes getting all the details wrapped up is like pulling teeth. The clean-slate attraction of starting a new job can overpower the drudgery of completing the old. This temptation can backfire, souring good clients and losing referrals.

Owner-builders doing their own work are often tempted to move into their house be-

10 **Finish the Job.** Resist the urge to move in before the finish work is complete.

fore the finish work is done, thinking it will be easier to do when it's close at hand. After a while, they don't notice the lack of trim, especially if there's furniture in the way, and it becomes harder and messier to set up the tools and work around the obstacles. It can be a strain on a marriage if the bathroom doors aren't hung after a few years of residence. I advise owner-builders to get everything done before they move in, and contractors to finish all work before they move on. They'll be glad they did.

Will Beemer is executive director of the Heartwood School in Washington, Massachusetts, and executive director of the Timber Framers' Guild of North America.

A Simple Approach to Raised-Panel Wainscot

■ BY GARY STRIEGLER

Raised-panel wainscot speaks of a time when craftsmen had an abundance of skill and the time to display their talents. But building traditional raised-panel wainscot is a complex, time-consuming process that few people can afford. I've simplified the process. I can build good-looking raised-panel wainscot in place using basic carpentry tools and a router table. The 12-ft. by 18-ft. room pictured here took me three days to complete.

All the materials I use to make raised-panel wainscot are readily available. The bolection molding that bridges the gap between the wainscot frame and the raised panel is a stock molding from White River Hardwoods/Woodworks Inc. (www.mouldings.com; 800-558-0119). Because the wainscot will be painted, I can use medium-density fiberboard (MDF) for the raised panels. MDF profiles well and is stabler and less expensive than solid wood.

Start with a Detailed Layout

Before I cut any wood, I snap horizontal chalklines to represent the top and bottom rails to see if the wainscot height is appropriate for the room. Next, I determine how many panels I need. They should be wider than they are tall, and to my eye, an odd number of panels looks best. For this project, five panels fit perfectly on the longest wall.

Pleasing Proportions from Top to Bottom

Although small in section, the moldings in this wainscot have a big impact on its overall appearance.

1 Bolection-molding cap   **2** Top rail   **3** Bolection molding

4 MDF raised panel   **5** Bottom rail   **6** Panel molding

7 MDF baseboard   **8** Cleat   **9** Shoe molding

The writing is on the wall. A full-size layout on the wall lets the author see the wainscot's proportions and serves as a guide for assembly.

I determine the panel width by subtracting the total width of the stiles (vertical pieces) from the length of the wall. I take into account that one panel will overlap another in corners and that the lapped stile should be ¾ in. (the thickness of the stile) wider. Then I divide the result by the number of panels. Once I have the panel width figured out, I use a level to lay out the stiles on the wall. If any electrical boxes fall on a stile or panel edge, I have my electrician move them.

Build the Longest Frame First

Because it's easier to make fitting adjustments to the smaller frame of the shorter wall, I assemble and install the longer frame first. The frame stiles and rails must be the same thickness, so my first frame-making step is to run all frame stock through a portable planer. I also use this machine to plane all rails to width by running them through the machine on edge.

Finding clean, straight lumber in 18-ft. lengths is nearly impossible, so I use pocket screws to assemble the top and bottom rails from two shorter pieces. I make certain that the top rail's butt joint breaks on a stud and is not in the same panel as the bottom rail joint. I make the overall rail lengths ⅛ in. shorter than the wall to ensure that the rails fit easily. Any gap will be covered by the adjoining wall's overlapping stile.

I cut all the stiles on a miter saw using a stop block for accurate repetitive cuts. Next, I lay out the lumber for the top and bottom rails against the long wall. I transfer the layout from the wall to the rails using a speed square. Then I lay the stiles along the rails' layout lines (see the photo above). (Remember, the lapped corner stiles are ¾ in. wider than the others.) To assemble the rails and stiles, I use a pocket-hole jig and connect the pieces with yellow glue and pocket screws (see the sidebar on the facing page). Glue should seep out around the joint and will be sanded smooth once it sets up.

Cleats Support The Frame

I nail a series of cleats along the wall to support the frame and raise it to the height of the chalkline I snapped earlier. The cleats

Top rail

1¼ in. pan-head screw

Pocket hole

Stile

Assemble the Frame Next to the Wall

Use a pocket-hole jig to drill the stiles for assembly. Lay out the rails and stiles along the wall, then glue and screw them together. Get some help lifting the frame into place because butt joints along the rails could flex. Cleats support the frame for nailing and serve as nailers for the baseboard.

are the same depth as the frame and also act as nailers for the baseboard that is installed later.

I need help to tip the 18-ft.-long frame into place. The frame isn't that heavy, but it's awkward for one person to lift into place. And although the pocket-screw joinery is extremely strong, the butt joints along the rails are the weakest point of the assembly and could flex if not supported properly.

Once the frame is up, I set it on the cleats so that I can nail it to the studs using a 15-ga. finish nailer. I then smooth the rail-and-stile joints with a 120-grit sanding disk in a random-orbit sander.

Raised Panels Are Shaped at the Router Table

The bolection molding that I chose for this wainscot projects 1½ in. into the frame's opening. To keep the raised-panel profile from being concealed, I size the panel to fit inside the frame and molding.

MDF Panels are Flat, Smooth, and Stable

Size, Raise, and Install the Panel

The author uses a gauge block to determine the panel size and mark its layout lines (top photo below left). Cutting the raised-panel profile in multiple passes on the router table ensures smooth results (middle photo below left). He then installs each panel by driving 2-in. brads through the flat panel edge (bottom photo below left).

Top rail

1⅛ in. gap

MDF panel

Because I knew the profile of the raised panel to be made, I determined that the bolection molding should overlap the raised panel by ⅜ in. This amount of overlap covers the thin, flat edge of the panel without hitting the raised profile, gives me enough wiggle room to adjust the bolection molding, and covers the brads used to fasten the raised panels to the wall.

I cut a 1⅛-in. gauge block (the 1½-in. molding projection minus the ⅜-in. overlap) from scrap lumber and use it to lay out the raised panel in the frame. I mark only the corners because they determine both the placement and the size of the MDF raised panels.

I make the panels out of ¾-in. MDF on a tablesaw. Because of the fine dust that is created when cutting MDF, it's necessary to wear a respirator, even in a well-ventilated area. After the rectangular panels are cut to size, I bring them to the router table to shape them into raised panels (see the middle photo on the facing page).

I don't rout the MDF in one pass; instead, I make three passes for each side. This process can be a bit time-consuming, but by setting the router depth once and increasing the cut by moving the fence to reveal more of the router bit with each pass, I get a smoother profile, keep the router from overheating, and reduce the chances that the panel will kick back. Plus, I don't have to duck under the table to readjust the router.

I work this phase like a production line. I rout four sides of every panel, increase the depth of cut, run every panel through again, and repeat the process. This approach ensures uniform panels cut to the same depth.

To install the panels, I simply rest the panel on a 1⅛-in.-thick gauge block along the bottom rail and position the panel along the layout lines I made earlier. Along the panel edges, I drive 2-in. brads into the studs. I can nail the MDF panels to the wall because they are extremely stable and won't move with swings in humidity.

Molding Bridges the Gap between the Raised Panel and Frame

The bolection molding has a rabbet on its back, allowing it to seat itself in the framed opening while covering any joints. To cut a piece of bolection molding properly, the rabbet is elevated by a sacrificial block the same depth as the rails and stiles; then it's just a matter of making 45 degree miters and fitting the molding to the opening.

After cutting the molding, I use a 23-ga. headless pinner to install each mitered frame

Mitered Molding Surrounds the Panels

Avoid Splits by Driving Pins Instead of Nails

A 23-ga. headless pinner won't split even the most delicate molding, provides sufficient holding power, and makes nearly invisible fastener holes. Drive the pins into the frame members and panel edges as shown in the drawing at right.

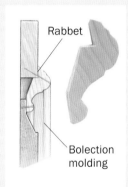

Rabbet

Bolection molding

A Flat Panel Avoids Busy Windows

Instead of crowding a raised panel under the window, this area gets a simple step-out treatment. I use a 3½-in.-wide window stool to accommodate the extra depth for this detail.

1. After two MDF nailing strips the thickness of the wainscot frame are installed, a single, flat MDF panel is built out 1½ in. under the window. I size the MDF panel to cover the adjoining wainscot frame by ¼ in. on each side.

2. Next, I install the MDF baseboard on top of the flat panel. At the step back, I use a mitered return that is fastened with yellow carpenter's glue and 23-ga. headless pins.

3. The 2-in. panel molding that caps the baseboard is nailed every 16 in. with 1½-in. brads.

by nailing the molding to the stiles, rails, and panels. I reinforce the miter joints with glue and cross-nail them to hold the joints tight.

Finish Up with Bolections and Baseboards

Once I've finished mitering the frames around the panels, I'm in the home stretch. To cap the wainscot, I use a rabbeted bolection molding that contains a chair-rail profile (inset photo, facing page). Brads that are driven into the top rail attach the molding, and the rabbet hides the joint between the cap and the wainscot. Because I use a backband on window and door casings that is

thicker than the bolection molding, a simple butt joint is all that is necessary to terminate the molding at the windows and doors.

I make the baseboard detail out of 6-in.-wide MDF topped with a 2-in. panel molding. The baseboard is nailed to the cleats and bottom rail, and the molding is fastened to the bottom rail. At the doorways, I clip the portion of the baseboard that stands proud of the casing with a 45 degree bevel to ease the transition between the two details.

Gary Striegler runs Striegler and Associates, a custom-home building company in Fayetteville, Arkansas. Photos by James Kidd.

A Cap and Baseboard Finish Off the Wainscot

Keep Installation Simple with a One-Piece Molding

A bolection molding with a rabbeted bottom edge caps the wainscot and contains a chair-rail profile. The molding is a simple alternative to the traditional two-piece chair rail. The author completes the wainscot with a 6-in.-wide MDF baseboard topped with a panel molding.

Brad

Rabbet

Crown Molding Around a Cathedral Ceiling

■ BY GARY M. KATZ

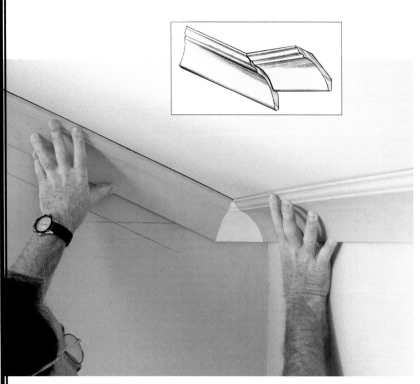

The problem and the solution. When crown molding runs up a vaulted ceiling, the profiles won't match at the corner (see the inset drawing). So instead of using different-size moldings, a transition piece turns the corner using the same profile (see the photo above).

For years, whenever people asked me to install crown on a cathedral ceiling, I'd just shake my head and say, "No, that always looks dumb," and hope they'd forget about it. The problem is that when crown goes from the horizontal run to the run up the rake of a vaulted ceiling, the intersecting faces don't match perfectly at the corner (see the inset drawing at left).

A couple of years back, I ran into Joe Fusco, a veteran finish carpenter and cabinetmaker, at a trade show. Joe showed me a solution using two different-size moldings. A larger version on the horizontal run lets you maintain the same spring line up the rake. But the same crown profile isn't always available in different sizes; and besides, the smaller crown on the rake looks a little funny to me.

So I was intrigued earlier this year when another colleague of mine, Mike Sloggatt, told me about cutting a transition piece that lets him make the turn at a vaulted corner using the same-size molding. I may not be a "master" carpenter, but I am a dogged one. So on my next job with a vaulted ceiling, I stayed in one corner until I figured it out.

Horizontal Runs Must Be Ripped

Crown molding is designed to work with a flat ceiling. So with a vaulted ceiling, horizontal runs of crown molding must be modified to fit against the slope of the ceiling.

For this modification I rip a little from the back side of the crown. To determine the rip, I place a piece of crown against the ceiling and scribe the angle (see the top left photo below). Then I build a simple fixture that clamps to my tablesaw fence and lets me rip the crown molding quickly and safely. The transition piece I mentioned also is made from this modified molding, so be sure to rip a little extra.

A Transition Piece Turns the Corner

The key to this approach lies in visualizing the transition piece as a short continuation of the horizontal run. The transition piece turns the corner with a 45-degree miter just like a normal piece of crown. But then another miter cut makes the transition to the angle of the ceiling or the rake.

To get the miter angle for that transition, you just read the ceiling angle, subtract 90 degrees for the other side of the transition, then subtract the result from 180 degrees and divide the remainder in half. Huh? Well, math isn't my best friend either, so I rely on a simpler method (see the drawing below).

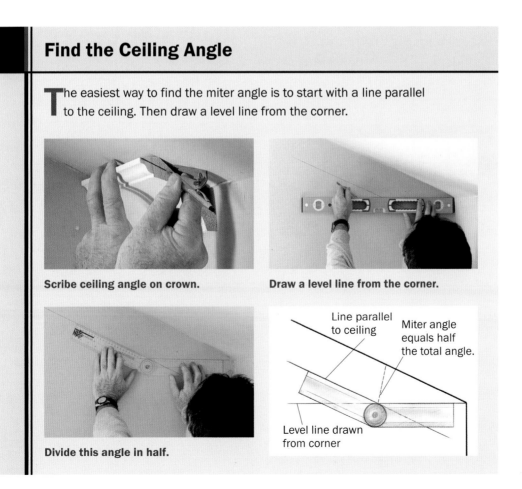

Find the Ceiling Angle

The easiest way to find the miter angle is to start with a line parallel to the ceiling. Then draw a level line from the corner.

Scribe ceiling angle on crown.

Draw a level line from the corner.

Divide this angle in half.

Line parallel to ceiling

Miter angle equals half the total angle.

Level line drawn from corner

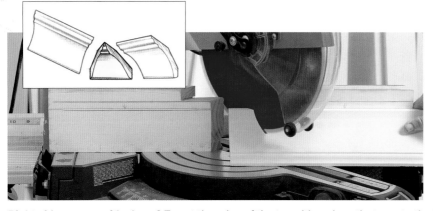
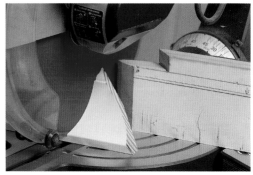

Right side up or upside down? To cut the edge of the transition piece that meets the horizontal run, place the crown upside down against the miter-saw fence (photo above). For the edge where the transition piece meets the run up the ceiling, flip the piece and cut the miter angle right side up (photo right).

Back-Cutting Crown Molding

Trick of the Trade

The quickest and safest way to rip the back cut on crown molding is with a shop-built fixture that clamps to a table-saw fence. The fixture consists of a vertical accessory fence with an angled table attached to it.

For the table I rip a 45-degree bevel on both edges of a piece of 1x and rip a short vertical shoulder on the lower edge. A stop, which is interrupted for the blade, attaches to this shoulder to keep the crown from sliding off. To rip this crown, I had to crank

my blade only to about 20 degrees, but the blade had to be raised quite high. So please be careful, and always use a push stick to complete the cut.

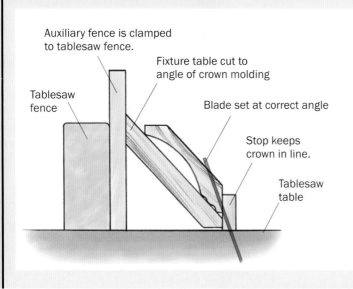

Auxiliary fence is clamped to tablesaw fence.

Fixture table cut to angle of crown molding

Tablesaw fence

Blade set at correct angle

Stop keeps crown in line.

Tablesaw table

First, I place a 1x6 board with its narrow edge against the ceiling and trace a line on the other side of the board. Then I draw an intersecting level line from the inside corner. A protractor on the intersection of these lines gives me the exact angle of the outside corner. I halve that angle for the miter angle.

The Cuts: Some Upside Down, Some Right Side Up

A basic carpentry rule is to cut crown molding upside down on a miter saw. But this rule applies only to horizontal runs of crown. Crown molding running up the vaulted part of the ceiling should be cut right side up.

Because the transition piece goes from the horizontal run to the vaulted run, it has to be flipped for the two different cuts (see the top photos). First I cut the 45-degree angle for the 90-degree inside corner from a piece of modified crown upside down in the saw. The next cut is for an outside corner, albeit a shallow one. So I flip the crown right side up, set the angle, and make the miter cut. I cut the transition piece slightly short so that the profiles of the horizontal run and the vaulted run meet at the top in the corner. The rest of the pieces for the vaulted run are cut right side up.

Gary M. Katz is a contributing editor to Fine Homebuilding.

Crown-Molding Fundamentals

■ BY CLAYTON DEKORNE

The first time I installed crown molding, I couldn't believe how easy it was. The builder I was working with knew exactly where every stud, truss chord, and piece of blocking in the condo complex was located. The walls were flat, the corners square, and the ceiling level; there were no outside corners and no extralong runs. I wouldn't find out that I'd been lucky rather than smart until a year later when I tried to put up crown in an old Victorian town house, where nothing was plumb, level, or square, and it seemed as if no framing existed beneath the plaster.

Angled between the ceiling and the wall, crown molding inhabits a three-dimensional space, making it one of the most demanding types of trim to install. When the wall turns a corner, crown makes two turns: One is along the ceiling plane, and one is along the walls, requiring compound cuts to join the pieces. Any discrepancy in either plane introduces a twist or bow that can alter the angles. Crown also is affected by shifting house movements acting on both planes, so it needs to be nailed securely to a stable surface to ensure that the joints don't separate over time.

Every Crown Gets Backing

Even when I'm following a first-rate framing crew, experience has taught me to check every corner with a framing square and to note any serious irregularities. (A gap between the square and the wall corner of even 1/8 in. can make it difficult to close the joint on a large piece of hardwood crown.) At the same time, I eyeball the wall and ceiling intersection, looking for any irregularities that might hinder the crown molding from making full contact. If I encounter a saggy ceiling or a wavy wall, I take the time to figure out if corrective measures are needed (see the sidebar on p. 166).

Even when everything is square and true, I still have to make arrangements for fastening the crown. Simply nailing into the studs and ceiling joists secures the crown only along the top and bottom edges. Inevitably, the edge splits, or the small finish nails don't find good purchase and give out after a few years of seasonal movement. I avoid such aggravations by installing a backer of rough framing lumber that allows me to nail the crown wherever I want.

For modest-size crown, my typical backer is a straight 2x4 ripped at a bevel to match

Two Tips to Make the Job Go Smoothly

Always Put Up a Backer

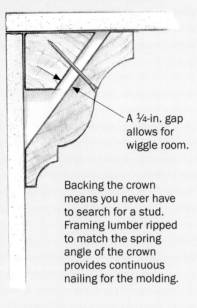

A ¼-in. gap allows for wiggle room.

Backing the crown means you never have to search for a stud. Framing lumber ripped to match the spring angle of the crown provides continuous nailing for the molding.

A framing square and a scrap of crown determine the backer dimensions. After ripping the material on a tablesaw, the author uses 3-in. screws to attach the backer to the framing.

Always Start with a Coped Joint

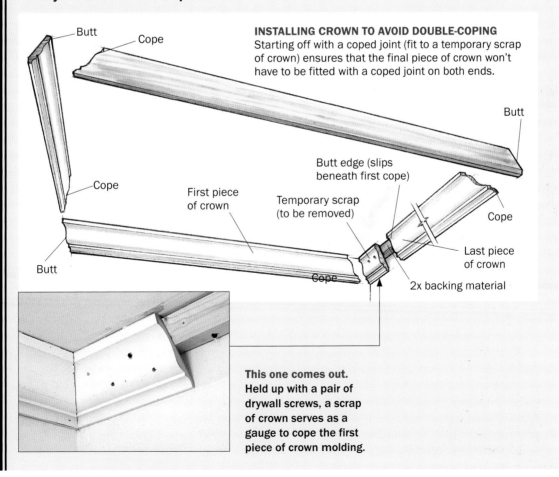

Butt

Cope

INSTALLING CROWN TO AVOID DOUBLE-COPING
Starting off with a coped joint (fit to a temporary scrap of crown) ensures that the final piece of crown won't have to be fitted with a coped joint on both ends.

Butt

Cope

First piece of crown

Butt edge (slips beneath first cope)

Temporary scrap (to be removed)

Cope

Last piece of crown

Butt

Cope

2x backing material

This one comes out. Held up with a pair of drywall screws, a scrap of crown serves as a gauge to cope the first piece of crown molding.

Coping Inside Corners

It's possible to assemble inside corners with a miter. However, even if the corner is perfectly square, most miter joints open after a few seasons of expansion and contraction. I prefer to cope crown molding. A coped joint is a marriage of opposites. On one side, the molding is cut square and butts tight to the wall. On the other side, the molding is "coped" to match the profile of its neighbor (see the photo at right).

A coped joint starts with a compound miter, the same as for an inside miter (see the left photo below). The contoured edge along the face of this cut defines the profile to be coped. I mark the edge of this profile with the flat of a pencil lead so that it's easier to see as I cut it with a coping saw (see the middle photo below). I always back-cut slightly by tilting the coping saw past 90 degrees to ensure that the two pieces of molding intersect only along the profile line. When cutting with a coping saw, I typically saw toward the curve from different directions so that I don't have to twist the saw and risk breaking the fragile edge of the profile (see the photo below right).

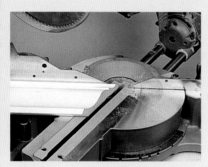

An inside-miter cut defines the profile of the cope.

Back-cut the cope, to ensure a tight joint.

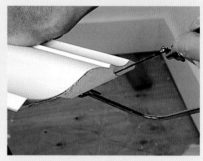

For best results, approach the cut from more than one direction.

the spring angle of the crown (see the top photos on p. 159). Rather than cut the backer to fill the space behind the crown completely, I lay it out and cut it so that the beveled face is about ¼ in. shy of the back of the crown. This gap allows enough wiggle room to fine-tune the joints if needed. The crown still firmly contacts the ceiling and wall at the edges, but its meaty belly is pinned with a stout finish nail.

I find it's usually easier to run the backer continuously. But if I'm running low on backer material or if I want to avoid high spots in the wall or ceiling, I may run the backer in sections—whatever it takes to give me nailing every 16 in. to 24 in. On walls long enough to require two lengths of crown, I figure out in advance where the joints will be and leave about a 16-in. gap between backer sections to accommodate the method I use to preassemble scarf joints (see the sidebar on p. 162).

Sliding two sticks against opposite walls and marking them after each makes contact is an effective method for measuring between walls. The ends of the pinch sticks are tapered to make one positive contact point.

Plan the Installation to Avoid Double-Coping

Before I start chopping up the molding, I scope out the installation order. To avoid coping two ends of the same piece of molding and to eliminate the time and fuss of resetting my saw for each cut, I try to set up all my copes on the same end of each piece. Most right-handers find it easier to cope right, butt left. The direction doesn't matter as much as staying consistent; find what suits you and stick to it.

If the room has an outside corner, I usually start from there. Otherwise, I start with the trim on the longest wall opposite the door. I install this piece with one coped end (see the sidebar on the facing page) and the other end butted square into the corner.

When I set up all the cuts in the same direction, the coped end of the first piece of the molding must be installed against a temporary block (see the photo on p. 158). This block is a scrap piece of crown, about a foot long, that is screwed temporarily to the backer. I nail the first piece of crown near the center and toward the butt end only. This placement gives me enough support to hold up the entire length of crown, but it leaves the coped end free so that I can slip the last full-length piece of crown in place of the starter block when I finally get there (see the drawing on p. 159).

Pinch Sticks Make for Easy, Accurate Measurements

Because joint compound tends to build up in the corners, it's hard to get an accurate measurement by bending a tape measure into a corner. For long lengths of crown, I add a heavy $1/16$ in. to all my measurements;

Outside Corners Are Simple Unless They Aren't Square

When I encounter an obtuse or acute angle corner, I use a couple of sticks and a bevel board to measure the exact miter angle. The bevel board is nothing more than a square piece of wood with some carefully drawn angled lines (from 0 degrees to 90 degrees). I made mine using a simple protractor and a sharp pencil (see the bottom right photo). To get the exact bevel angle for an out-of-square corner, start with two short scraps of wood that overlap at the corner (see the bottom middle photo):

1. Mark the long point of the angle on the upper piece.
2. Mark the short point on the same piece.
3. Find the angle on the marked piece with a bevel square.
4. Then lay the bevel square on the bevel board and read the angle.

A more modern alternative is the Bosch® DWM40L Miterfinder (see the bottom left photo). This ingenious device can measure any angle to within one-tenth of a degree; then with the push of a button, it gives you the miter and bevel angles for any crown molding.

Marking the long point is generally more accurate than marking the short point. To establish the long point of an outside miter, trace the top edge of a scrap of crown onto the ceiling, on both sides of the corner, then measure to the intersection of the two.

The miter formed by the intersection of two boards is transferred to a bevel square.

A homemade bevel board reveals the angle of the miter.

For about $110, a digital protractor does the thinking for you.

then I put a slight back-cut on the square end. That way, even if the piece of crown is slightly long, it can be sprung into place, and the point at the crown's bottom edge digs into the wall.

To measure for short lengths of crown, I always use pinch sticks, butting one end to the previously installed length and the other to the opposite wall (see the photo on p. 161). By holding the two sticks together and sliding them until one end of each hits the wall, I get an accurate representation of the distance that can be transferred easily to the molding. Once I have the distance, I mark the sticks with two lines for positive realignment, and I use a spring clamp to hold the sticks together when transferring the length to the molding.

Outside Miters Must Fit Precisely

If the corner is crisp and square, I measure the length for an outside miter by holding the stock in place and marking the bottom edge of the molding where it meets the outside corner. (Bear in mind that drywall corner bead pushes the corner out slightly, so you may have to back-plane the miter to get a tight fit.) Plaster corners may be rounded slightly, making it hard to measure the exact length along the crown's bottom edge. If the walls are reasonably square to each other (and only the corner is rounded), I mark the ceiling by tracing along the top edge of a piece of molding, first along one side of the corner, then along the other. Where these lines cross on the ceiling marks the long point of each miter cut (see the top photo on the facing page).

If you have to assemble more than one outside corner, consider cutting them all at once. I still try to map my cuts so that all the cuts for the copes angle the same direction. Rooms with bump-outs often have a few small pieces of crown. Wherever a small piece must be coped on one end and mitered on the other, I cope the piece before

I cut it to length. In this case, I might have to tack up the coped length of the outside corner temporarily so that it can be removed later to fit the last piece of crown that laps behind it.

Dual Compound-Miter Saw Makes Cutting Crown Less Confusing

If there's any secret to cutting crown, it's simply this: Keep track of which edge of the molding is the bottom and which edge is the top. Doing so eliminates a lot of head-scratching when you're standing at the saw trying to remember how the molding will sit when actually nailed into place. I try to group similar cuts together. That is, I first cut all the scarf joints, then cut all the outside-miter joints, then move on to the inside-corner cuts. This strategy saves time with any trim. But with crown molding, it's especially helpful because it lets me keep track of the three-dimensional orientation of the molding.

The methods you use to cut inside and outside corners depend on the type of saw you're using. With a conventional miter saw—and narrow crown—I use a simple jig that holds the crown in the saw at the precise angle in which it will be installed.

Unless you've got a huge miter saw, such as the 15-in. Hitachi C15FB, you need to use a sliding compound-miter saw if the crown is wider than 5½ in. A sliding compound-miter saw allows you to cut the crown flat on the saw table. With most sliding miter saws, the motor head pivots in only one direction; this means that you have to line up either the top or the bottom edge of the molding on the fence to get the cut in the right direction, so you wind up flipping material a lot.

If there's any secret to cutting crown, it's this: Keep track of which edge is the bottom and which edge is the top.

What to Do When the Wall Is Longer Than the Stock

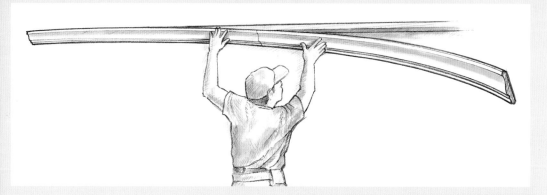

If a wall is longer than the crown stock available, you need to splice (or scarf) two lengths together. Rather than attempt to fashion a good fit on the wall, I create my scarf joints first.

For scarf joints, I depend on two critical devices: a biscuit and a splint. I begin by cutting a 20-degree bevel and a 20-degree miter—or 20/20 compound cut—on one end, and a corresponding 20/20 compound cut on the mating piece. I prefer a compound cut because it's easier to hide the joint; to make this cut on a standard miter saw, use the jig as shown in the top photo on the facing page. With the molding and the biscuit joiner lying flat on the same surface, I make biscuit slots in each piece. (1)

The slot should fall in the center thickness, not through the contoured face of the molding. The long point of the bevel on the compound cut holds the biscuit joiner away from the surface in which the slot is cut, but by setting the biscuit joiner to a maximum depth, I can fit a smaller #0 biscuit into the slot.

To fasten the two pieces more securely, I cut a strip of ¼-in. thick lauan to act as a splint across the back and to bridge the joints. I fasten the splint with plenty of glue, using ½-in. staples and a standard staple gun to hold the strip until the glue cures. (2) After securing the splint to the scarf joint, I smooth out the joint with sandpaper. (3) Until they're securely fastened in place, I make sure to handle these scarfed pieces with extra care; they can snap in two easily.

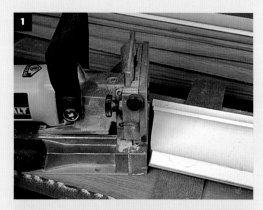

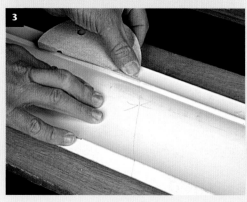

Build a Jig or Buy a New Saw?

If you have to cut crown on a standard miter box, an auxiliary plywood fence with a stop applied to the front edge ensures that the molding is held at the correct angle in relation to the blade (see the bottom photo). Because the saw head on a dual compound saw tilts in two directions, you never have to flip the molding around to set up the correct cutting angle, which is a big advantage when working in tight or crowded quarters (see the top photo).

Saggy Ceilings and Wavy Walls Demand Special Attention

In old houses, where walls and ceilings aren't smooth, I begin by mapping the planes with a level and string to find the low points on the ceiling. Long lengths of crown are surprisingly flexible and may move with gentle bumps and shallow dips. Most of the time, however, some remedy is required.

One way to solve this problem is to scribe the top edge of the molding to the ceiling. When scribing, I don't want to bite too deeply into the top edge, or the scribe will be easily visible. Generally, I don't remove more than a third of the top reveal.

Another fix is to skim-coat the wall or ceiling surface with plaster or joint compound to conceal gaps after the crown has been installed. In this case, I install the crown in line with the high spots. To float the mud, I make a screed from a short piece of mangled crown molding that has a notch in it that keys with the edge of the installed crown molding (see the photo at right). This screed allows me to apply the mud to a depth that precisely covers the gap. I usually make the screed about 18 in. long, something that gives me a long, thin taper of mud that seamlessly feathers into the existing surface.

Homemade screed levels the peaks and valleys. Rather than filling gaps in wavy walls with great gobs of caulk, a notched screed, made from a scrap of crown molding, enables the drywall finisher to feather those gaps back into the wall, where they belong.

Recently, I began using one of the new dual compound-miter saws. I admit I was skeptical at first that such a big lug of a saw would offer much of an advantage—unless I was dicing up huge pieces of lumber. But a dual compound-miter saw tilts in two directions, so you never have to flip the molding around. This is a big plus in a tight workspace.

With the dual compound-miter saw, I always keep one edge against the fence and move the saw head to the left or right depending on whether I am cutting an inside or outside miter, left or right.

With any compound-miter saw, you still need to find the correct miter and bevel settings. These settings change, depending on the spring angle of the crown molding you're installing, and there's no obvious correspondence between all the angles. I always keep handy a chart that lists the angle settings for the most common spring angles. For moldings that aren't included on the chart, I use the Bosch Miterfinder (www.boschtools.com; 773-286-7330) to calculate the angle settings, or I simply make a reasoned guess and cut a couple of test pieces until I get the angles right. Then I write those numbers down so that I won't forget them.

Clayton DeKorne is a carpenter and a writer in Burlington, Vermont. He is the author of Trim Carpentry and Built-ins *(The Taunton Press, 2002), and the co-author of* For Pros By Pros: Finish Carpentry *(The Taunton Press, 2007).*

Building a Fireplace Mantel

■ BY GARY KATZ

Mantel built in three sections. A 2x4 prop (left photo) holds the mantel top level while the author nails a pilaster to the panel wall. The three components—two pilasters and a mantel top—were built off-site; the texture on the oak comes from scraping out the soft grain.

At first glance, fireplace mantels seem impressively intricate and outrageously expensive, but frequently they aren't. I know because a good client asked me to build a copy of a mantel from his previous home. After seeing the original mantel and identifying its parts, I was able to build the new mantel simply and economically using solid stock, plywood and manufactured moldings (see the photo above). Even though I've been a trim carpenter for many years, anyone with a basic knowledge of woodworking can build a fireplace mantel with the techniques I used here.

Strips of MDF Simplify Flute Spacing

For this mantel, the first things to build are the pieces flanking the firebox, or the pilasters. I make each pilaster with three pieces—two sides and one face—mitered together along their length. I rip the faces and sides from solid oak and cut them slightly longer than their finished length. Once the pilasters are built, I'll trim them to size on a radial-arm saw.

Side View Shows Combination of Plywood and Solid Wood

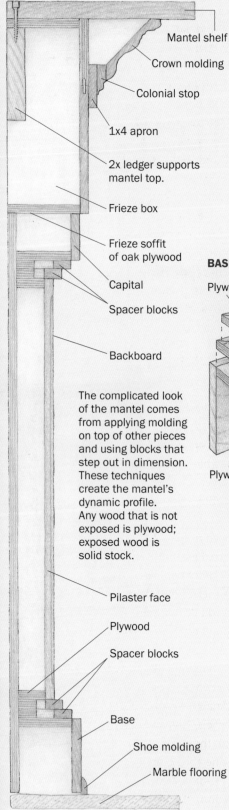

Mantel shelf

Crown molding

Colonial stop

1x4 apron

2x ledger supports mantel top.

Frieze box

Frieze soffit of oak plywood

Capital

Spacer blocks

Backboard

The complicated look of the mantel comes from applying molding on top of other pieces and using blocks that step out in dimension. These techniques create the mantel's dynamic profile. Any wood that is not exposed is plywood; exposed wood is solid stock.

Pilaster face

Plywood

Spacer blocks

Base

Shoe molding

Marble flooring

Cut multiple flutes without moving the fence. The pilaster faces were fluted on a router table by registering them against 1-in. MDF strips instead of directly against the rip fence. After each pass, a strip is removed; the rip fence is not moved.

BASE IS A SERIES OF BLOCKS

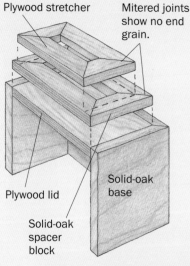

Plywood stretcher

Mitered joints show no end grain.

Plywood lid

Solid-oak base

Solid-oak spacer block

Joining the pilaster to the capital. After building up the capital with two 1-in. oak spacer blocks, the pilaster's plywood end is screwed to the capital. All of the blocks are flush.

Use plywood where it won't show. Nailed to the pilaster, a backboard eliminates the need to scribe fit the fireplace's marble veneer to the pilaster profile. For economy's sake, the backboard consists of two pieces of solid oak on the exposed ends and plywood in the hidden center, all glued and biscuited together.

Although I rip the sides of the pilasters to width with one edge square and the other beveled 45 degrees, I rip the faces slightly wide and with square edges. Later, I'll bevel the edges at 45 degrees for mitered joints, but right now I'd rather push a square edge along the tablesaw's rip fence. The long point of a bevel can jam under the fence, and in this case I'll be pushing the face stock through the fence six times to make parallel decorative grooves, or flutes.

Although fluting looks ornate, it isn't difficult to accomplish with a table-mounted router. My Makita portable 8-in. tablesaw is designed with a router mount underneath, so the bit sticks up through an opening in the table. The fluting also can be cut by running a router along a straightedge or by using a router guide. While these methods might be safer, they are slower and less accurate than a table-mounted router.

I use a ½-in. round-nose (corebox) bit to cut the flutes, set to a depth of ¼ in. To make the process easy, I place multiple strips of medium-density fiberboard (MDF) between the rip fence and the workpiece (see the sidebar on the facing page) and remove one strip after each pass, making it unnecessary to adjust the rip fence for each flute.

For the fluting on these 6-in.-wide pilasters, I make the MDF strips 1-in. wide: ½ in. for the flute plus ½ in. for the space between flutes, which gives me even spacing across the face. The MDF strips stay put because they butt into a stop block that's screwed to the table extension. The stop block is positioned so that the workpiece bumps against it, resulting in flutes that end 1¼ in. from the bottom of each face.

On the MDF strips, I draw a line 1¼ in. beyond the feed side of the router bit; this line indicates the beginning of the fluting. I carefully lower the workpiece past the line for the initial plunge and slowly pull it back until the end of the stock lines up with the 1¼-in. mark across the MDF strips. Then I feed the board through until it butts into the stop block, remove the workpiece and an

MDF strip, and repeat the process five more times.

After fluting the faces, I rip the edges at 45 degrees. Then I spread glue on the miters of the side pieces and the face, hold the joints tight with my fingers, and nail the miters together. Plywood or MDF backer blocks nailed inside the pilasters at each end support the joints (see the right drawing on the facing page); the backer blocks also serve as a backing both for the bases and for the capitals, which are the decorative blocks that will adorn the bottoms and the tops of the pilasters.

It's important to close up the miters as tightly as possible while the glue is still wet. I use a scrap of hardwood or my hammer handle to roll the sharp edge down just slightly, which flattens the mitered edge and closes the joint. Fine sandpaper finishes this process, leaving the mitered edges softly eased.

Making Capitals and Bases

The pilasters have a capital and a base; each capital and base has two 1-in. spacer blocks (see the left drawing on the facing page). The bases, capitals and spacer blocks are solid oak, and each one is larger than the one that it's attached to, creating a profile of ¾-in. steps.

The spacer blocks are ripped from 1x2 oak and are mitered on a chopsaw. I make them with four sides, like picture frames, with all miters glued and nailed together. I miter the corners because this mantel will be stained rather than painted, and I don't want any end grain to show.

The capitals and bases are built just like the pilasters, with a face and two sides mitered together. I use my tablesaw and rip fence to cut the 45-degree bevels, glue and nail these joints together, roll the edges, and sand them. I install a lid at one end using any material handy, usually ¾-in. plywood. The lids are backing for attaching the spacer blocks, and they aren't exposed.

Backboard Eliminates Scribing

On this job, a marble-veneer fireplace surround will be applied after I install the mantel. To make the mason's job easier, I mount the pilasters on a backboard (see the bottom photo on p. 168). This board provides a straight line for the marble installer to butt the material to; otherwise the installer would have to cut that marble to fit the profile of the pilasters. With the pilasters attached to the backboard, the installer has an easier job.

Because most of the backboard is hidden behind the pilaster (there's a ¾-in. reveal beyond the capital and base), there's no reason to make the whole backboard of solid stock. I use two pieces of 1x4 oak for the area that won't be covered by the pilasters and a plywood filler in the center. I join these three pieces with biscuits to make a 12-in.-wide backboard. With the pilaster covering most of the backboard, none of the joints will be visible; the joints just have to be flush to avoid gaps between the pilaster and the backboard. I make the backboards longer than the pilasters. Once the pieces are assembled, I'll trim the backboards flush with a circular saw.

Assembling the Pilasters

Now it's time to put all the pieces together. I glue up the first spacer block and center it on a capital or base so that the reveals are an even ¾ in. The pieces must be flush at the back, or else there will be gaps between them and the backboard. I nail the first block in place and position the second smaller block in the same manner. Then I attach a capital and a base to each pilaster, running screws through the backer blocks to draw the first spacer block tight (see the middle photo on p. 168). I also nail the backboard to the pilaster so that there's a ¾-in. reveal at the capitals and the bases.

The Frieze Is Just a Big Box

I build the top of the mantel starting with the frieze box that will sit above the capitals and support the mantel shelf. In this case I copied an existing frieze box, though usually I rely on a scale drawing or a mock-up of the mantel to determine the frieze dimensions. This frieze box overhangs the capitals by ¾ in. both in the front and on the sides.

Although the frieze box is 15½-in. high, I use 10-in.-wide solid oak only for the exposed portion of the box, the part below a 1x4 oak band, or apron. Over this apron I apply colonial doorstop and a wide crown molding, both stock items. The upper portion of the frieze box, therefore, is covered with trim, so I make the upper portion of the box from ¾-in. plywood and biscuit it to the solid oak (see the photo at left).

Building the frieze box is just like building the capitals and bases in that the front and sides are mitered, glued, and nailed, except I install blocks across the biscuit joints to strengthen the connections. I also nail in a couple of midspan blocks that provide backing for the bottom of the box, or the soffit, and the backing for the lid, both of which sit within the frieze box. The soffit

Frieze box has plywood upper half and solid-wood lower half. This part of the mantel top will be capped with a shelf and covered with trim, and only the solid-oak lower half will be exposed. The plywood and the solid wood are biscuited together, and blocking supports the joints.

Fitting the colonial stop.
The joint between the frieze box's solid-oak lower half and its plywood upper half is covered by a 1x4 oak apron to which is affixed the colonial stop.

will be exposed, so here I use oak-veneer plywood. The lid of the box can be made of anything. Its only purpose is to provide backing to secure the mantel shelf.

When installing the backing blocks, I take care to hold them 1½ in. inside the back of the box so that I can slip the box over the 1½-in. mounting ledger and have it fit snug to the wall.

On the assembled frieze box, I draw pencil lines showing the positions of the crown molding, colonial stop, and 1x4 apron. I place a short piece of crown on the frieze box with the top of the crown butted against a framing square as if it were the underside of the mantel shelf and scribe a line along the bottom of the crown. From that mark I locate and draw lines for the colonial stop and for the 1x4 apron. I cut and install the apron and the stop, mitering and gluing the outside corners (see the photo above). The crown stays off until the shelf is installed.

Putting on the Shelf

I make the mantel shelf from a single piece of 6/4 solid oak. The owner wanted the end grain to show, so I didn't miter the ends. On this job, the shelf overhangs the top of the crown molding 3 in. That overhang is consistent around the front of the mantel, too, so the width of the shelf is equal to the frieze box, plus the coverage of the crown, plus a 3-in. overhang. After ripping, crosscutting, and sanding the shelf, I use a ³⁄₁₆-in. roundover bit in my router to ease the edges.

To secure the mantel shelf to the frieze box, I place the shelf upside down on my workbench and position the frieze box on the shelf, centered and flush at the back. Then I screw through the frieze lid into the mantel shelf.

Installing the Crown Molding

With the shelf in place, I now can install the crown molding (see the photo below), which goes under the shelf and flush with the back of the frieze box. The main concern is getting the crown's outside miters tight.

I place the front piece of crown upside down on my compound-miter saw as if the saw table were the mantel shelf and the saw fence the frieze, and cut one end at a 45-degree miter. I tack this piece in place on the colonial stop, then cut the return piece. I'm no mathematician, but I am fair at trial and error, and that's the method I use to find the perfect angle on a compound-miter saw to cut crown molding. I cut scraps until they fit just right to determine the setting for the saw. Wish it were easier.

Fastening the crown.
Once the shelf is fastened to the frieze box, the crown molding is applied. The front piece goes on first, and the mitered ends are glued and nailed in place.

Masking tape makes the scribe line visible. Although the pilaster's backboard eliminates the need to scribe the pilaster to the wall, the capital must be scribed to fit beneath the frieze soffit. The pilaster is shimmed up until it touches the soffit; then it's scribed and cut. Masking tape makes the scribe line easier to see on the finished capital.

Once the saw is set, I cut the right side of the front piece first, then I hold it up to the colonial stop, where the crown eventually will be nailed. I check the fit on the right end with a short scrap, also mitered, then mark the bottom back edge of the left end. This is the ideal spot to guide in the blade of my Hitachi sliding compound-miter saw.

Once the miters are tight, I cut the backs of the return pieces so that they are flush with the back of the frieze box. Then I glue the mitered ends and securely nail all three pieces both to the stop and to the shelf.

Mantel Top Hangs on Ledger

The mantel top is mounted on the wall at the combined heights of the pilasters, the frieze box, and the shelf, plus 1½ in. of clearance for the marble flooring. I scribe this height on the wall and measure down from the mark the thickness of the mantel shelf and the frieze lid combined, which is 2¼ in. With a straightedge I scribe a line at this elevation across the wall. This line locates the top of the ledger that supports the mantel top. I rip the ledger from a piece of 2x8 and crosscut it ½ in. shy of the inside dimension of the mantel shelf, allowing me

a little room to adjust the final location of the mantel top. Then I fasten the ledger to the studs with 3½-in. screws and panel adhesive. On a masonry fireplace I would use plastic anchors and panel adhesive.

The frieze box slides over the ledger, and a couple of 2x4s temporarily support the box while the shelf, the soffit and the sides are scribed to the wall. The mantel top is prefinished, so I stick masking tape along the edges and scribe the lines on the tape, which makes them easy to see. I cut the top with a worm-drive circular saw. Because I cut all the scribe lines at a slight bevel, the bottom of the cut won't jut out and cause gaps to be visible between the mantel top and the wall.

Once the mantel top fits perfectly tight to the wall, I countersink holes and screw through the mantel shelf into the ledger. I use an inexpensive plug cutter by Vermont American (PO Box 340, Lincolnton, NC 28093; 704-735-7464) and make oak plugs to cover the screw holes, then sand them smooth for the finisher to touch up.

Installing the Pilasters

Because the pilasters are mounted on backboards, they only need to be scribed to the frieze soffit (see the photo above), not to the wall. If there were no backboard, I would install a cleat on the wall, much like the mantel-top ledger, and after scribing the pilaster to the wall, nail the pilaster to the cleat. In this case I don't need a cleat. I spread panel adhesive on the backboard and nail it right to the wall. I hold the pilasters 1½ in. above the floor, leaving plenty of room for the marble flooring to slide under the base. The resulting ¼-in. gap leaves an excuse to return to tack on some prefinished shoe molding and to take a picture (see the inset photo on p. 167) of the finished mantel for my portfolio.

Gary Katz is a carpenter and a writer in Reseda, California.

Installing Vinyl-Clad Windows

■ BY RICK ARNOLD AND MIKE GUERTIN

There are many significant milestones during a home-building project: first tree to fall, first scoop of earth, final roof rafter set, etc. But one of our favorites is window-installation day. For us, it's usually the day the bank designates the house as a "weathertight shell" and then issues the corresponding check. But even with that added incentive, we never rush the process of installing windows. The last thing a builder needs is a callback to fix a leaky window.

Check the Flanges before Installation

The windows we install most often are vinyl-clad, meaning they have a wooden frame wrapped with a protective layer of vinyl. Depending on the manufacturer and on the type of window, vinyl-clad windows have either an integral flange that is part of the extrusion covering the wood frame of the window or a flange that friction-fits into a slot on the head and sides of the window frame. The vinyl-clad double-hung windows we installed on this project were the slotted type.

Head flange slips under the house-wrap. As the window is lifted into position, the outside crew member guides the head flange under the housewrap at the top of the opening.

Make way for the flanges. When the window opening is cut in the housewrap, 2-in. slits are made beside the opening for the window flanges to slide through.

The windows may be shipped without the flanges attached, or the flanges may have come loose during shipping. So we check all the windows to make sure the flanges are attached properly before we begin installation. To reattach the flanges, we first press them into the slot by hand. Once they're in position, we drive them home using a 1x block set against the inward lip of the flange to cushion our hammer blows.

Flanges Tuck under the Housewrap

We usually start thinking about window installation before the excavator takes that first scoop of earth for the foundation. As the windows are chosen for each location in the house, model numbers and rough-opening sizes are noted. All these numbers are double-checked as the house is being framed.

We use housewrap as a secondary drainage plane. So instead of cutting the window openings diagonally from the upper corners, we cut straight across the top of the opening.

The bottom and sides of the housewrap are cut and stapled into the sides of the framed opening in the usual fashion. But we leave the top loose with no staples for at least 2 in. above the opening where the head flange of the window tucks under the wrap. Next, we extend the top cut about 2 in. out from the opening on each side for the side flanges to slide through (see the bottom photo on p. 173). As the housewrap is cut and stapled to the opening, we also check for protruding nails or anything else that could interfere with installation.

Two Crew Members Make Installation Easier

We've found that the quickest, safest way to install windows is with two crew members, one inside and one outside the house. Inside duties include handing the window out the opening, centering it, and adjusting it in the opening as the crew member on the outside directs. The outside crew member levels, plumbs, and squares the window and then fastens it to the wall.

Before lifting a window into position, we like to unlock and raise the lower sash to give us a better handhold. The crew member on the inside then passes the window unit out through the opening top first, angling it slightly to clear the opening. Having both sashes up makes the unit top-heavy, so we take extra care handling it.

As the window is lifted into the rough opening, the outside crew member tips the top toward the house and guides the head flange under the housewrap with the side flanges passing through the two slits made earlier (see the top photo on p. 173). Once the head flange is under the wrap, the window is lifted up and in. For the lift, the inside crew member, who must be on sure footing, is responsible for the weight of the window, while the outside crew member balances it and guides it into position.

Don't Let Go of That Window

Once the bottom of the window is pulled into the opening and is resting on the rough sill, the inside crew member hands the level out and gets ready to adjust the window in the opening. The weight of the window makes it want to fall out of the opening, so whenever one of the crew members needs to release his hold to grab a tool or to make an adjustment, he lets his partner know. That way, the window is never left unsupported.

Center the window first. The outside crew member holds the window in place while the window is centered in its opening from the inside.

At this point, the inside crew member centers the window side to side in the opening with a flat bar (see the photo above). The outside crew member then sets the level on top of the window (see the top photo at right) and directs the crew member inside to lift the low side with the flat bar until the unit is level. A shim shingle is then slipped between the bottom of the jamb and the rough sill to keep the window level until the outside can be nailed at the tops of the side flanges (see the bottom photo at right).

Every window we've installed has called for either 1¾-in. or 2-in. galvanized roofing nails or other broad-head nails to be used as fasteners. After the upper corners of the side flanges have been nailed (see the left photo on p. 176), we recheck the level to make sure nothing drifted.

Two sides of window installation. A level is held outside on the head of the window (above), and instructions are given to the inside crew member to shim the window as needed (left).

Top corners are first to be nailed. When the window is centered and the head is leveled, the nails are driven in the top corners to hold the top of the window in place.

Plumb jamb. No, not a breakfast spread. The inside crew member moves the window side to side with a flat bar as instructed by the outside crew member who holds a level against the jambs.

Check for square before nailing off the bottom corners. A measuring tape is stretched from corner to corner, and diagonal measurements are taken to make sure the window is square in its opening.

We intentionally avoid nailing the top flange at this point because with its friction fit into the head jamb, the flange could pull out while we're adjusting the sides for plumb in the next operation. Windows with integral flanges can be nailed at the top left and right corners without a problem.

Get the Sides Plumb and Square

Next, the inside crew member adjusts the bottom of the window side to side with a flat bar, while outside, a level is held against the window jamb (see the center photo above).

When the jamb reads plumb, the outside crew member drives a single nail at the bot-tom of the side flange without driving the nail home. The other jamb is then checked for plumb, and another nail is driven to secure the last corner of the window.

Before driving the two bottom nails home, we measure the window diagonally from corner to corner to make sure it's square (see the right photo above). Leveling the top and plumbing the sides of the window should make it square, but the diagonal measurements are the best confirmation that the window is square. Once we're satisfied, we set the bottom nails.

We straighten the side jambs by using a straightedge (see the left photo on the facing page) or by checking the reveal between the sash and the jamb by eye. The middles of the jambs are often bowed in or out and

A straightedge ensures smooth window operation. After the corners of the window are nailed in, a straightedge is placed against the jamb (left), and a nail is driven to hold the jamb straight (above). A straight jamb is essential for the sash to slide up and down properly.

Installing Windows for Rough Weather

Here in Rhode Island, ocean-front properties are in big demand, so we regularly find ourselves working within a stone's throw of the ocean. During the frequent storms that buffet the coast, rain rarely falls down from the sky; it blows in sideways.

In this environment, standard weatherproofing details do little more than channel wind-driven water into building assemblies. And leaks are most likely to occur around the windows.

The first thing we learned is to stop relying on caulk. Sun, wind, and water all conspire to penetrate even 50-year sealants in just a few months. Instead, we assume that wind-driven water will find its way in, despite our best efforts. Our strategy, then, is to make sure any water that makes it past our defenses is directed back to the outside.

Slope the Rough Sill

The details to redirect leaks occur before the windows arrive. First, we oversize the rough-opening height by ¼ in. during framing and install housewrap as we do for a standard installation.

At the bottom of the opening, we install a piece of 6-in.-wide clapboard with the thick edge toward the inside of the wall (see photo 1). The clapboard gives us a sill with a slight pitch.

We make the window drainage pan of adhesive-backed bituminous membrane (see photo 2). We cut a 9-in.-wide strip of membrane about 1 ft. longer than the width of the opening. The membrane is pressed onto the clapboard, with 6 in. or so of membrane run up the inside of each jack stud.

Next, we make a diagonal cut in the membrane starting ½ in. out from the bottom corners to create two flaps. But before we fold the

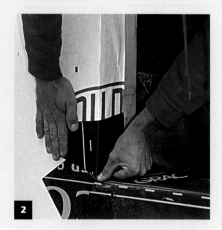

have to be straight for the window to slide in the jamb easily without being loose. After the jambs are straightened and secured with a nail (see the right photo on p. 177), the inside crew member checks the sash operation, then lowers and locks the bottom sash and checks the margins.

By the way, some window companies ship their windows with a plastic strap to keep the jambs from spreading during installation. We leave those straps in place until the jambs are shimmed and insulated.

Don't Skimp on the Nails

The process to this point takes about 10 minutes. Finishing installation typically takes another 10 minutes. But rather than have the inside crew member waiting around all that time, both crew members usually move on to the next window until all the windows are secured in place. Then one or both crew members can nail off all the windows.

All the window brands we use have nail holes prepunched in the flanges. The recommended nail spacing varies, so we check each manufacturer's instructions before nailing off. This brand calls for nails every flaps onto the walls, we press 6-in. by 6-in. pieces of membrane onto the outside wall to span the corners between the membrane flaps. We stick the filler pieces onto the wall so that the pan material overlaps onto them to form a leakproof corner.

The bituminous membrane is pressed tightly into all the corners to prevent a void that could puncture during window installation. On cold days, the membrane may not stick well to the sheathing or to the housewrap, but a few staples will hold it in place until the adhesive activates.

Seal the Flanges with Bituminous Membrane

We install the window as we do in a standard installation, except that we slit the housewrap 6 in. at the top corners rather than 2 in. Instead of using tape to seal window to housewrap, we seal the sides of the window to the housewrap with 6-in.-wide strips of membrane. The side strips extend through the slits and then stick to the wall sheathing under the housewrap.

For the head of the window, we slip a 6-in. strip of membrane under the housewrap (see photo 3). The head strip covers the top window flange and extends over the side strips. We let the side strips extend down over the pan flashing. As added insurance, we tape the housewrap to the membrane along the head of the window and out beyond the horizontal slit (see photo 4).

The bottom flap is again held in place by the siding so that any water that reaches the pan can drain out. By the way, this window-installation detail works great under vinyl siding. Vinyl siding is leakier than other sidings, so it can use some help to keep water out of walls.

6 in. to 8 in. Their holes are punched 4 in. apart, so we drive nails at every other hole.

When nailing off the head flange, we lift the flap of housewrap rather than nailing through it (see the left photo on p. 180). We always take special care when nailing off vinyl flanges, especially in cold weather when they can become brittle and crack easily.

On occasion, we've tried speeding up the nailing process by using pneumatic coil-roofing nailers. We're usually pretty good at hitting the prepunched holes. But missing with the pneumatic nailer can cause flange cracks, so we prefer to spend the extra time hand-nailing the flanges.

Most window manufacturers wrap flanges all the way around their windows. However, this brand has a soft, flexible flap at the bottom rather than a rigid flange. We tack down this flap with a couple of staples.

Housewrap Tape Seals the Windows

Window installation isn't complete until we've taped the window flanges to the housewrap. Taping the flanges helps prevent rainwater from entering the wall cavity, and it draftproofs the building envelope.

We first tape the sides where the flanges are on top of the housewrap. We extend the tape beyond the slits that we made for the

Tape air-seals the window. As a final measure, the flanges are taped to the housewrap. At the head of the window, the housewrap is taped to the window flange (top). The flexible flange at the bottom of the window is taped to the housewrap (bottom).

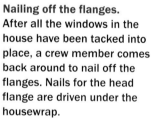

Nailing off the flanges. After all the windows in the house have been tacked into place, a crew member comes back around to nail off the flanges. Nails for the head flange are driven under the housewrap.

top flange and then run tape across the head of the window where the flange is underneath the housewrap (see the top right photo above).

In the past, we've treated the bottom flange two different ways. If the installation includes the drainage pan that we use in rough-weather situations (see the sidebar on pp. 178–179), then we omit the tape to permit any water that might get behind the window flanges to drain out. But for our standard installation, we tape the bottom flange to the housewrap to complete the air seal around the window (see the bottom right photo above).

Notes on Other Flanged Windows

Installation of casement and awning-flanged windows is nearly identical to the process we use for double hungs. The biggest difference is that we don't open the windows during handling. With the sash open, they become awkward, and their jambs tend to rack.

Another installation difference is that we boost the window units off the sills with thin (3⁄16 in. to 1⁄4 in.) blocks under the corners. These blocks give us space between the unit and the rough framing at the bottom to spray in air-sealing foam later. Other than that, the processes are identical.

Rick Arnold and Mike Guertin, contributing editors to Fine Homebuilding, *are builders and construction consultants in North Kingstown and East Greenwich, Rhode Island. They are the authors of* Precision Framing *(The Taunton Press, 2001).*

New Window in an Old Wall

■ BY RICK ARNOLD

A friend of mine recently had his appendix removed. Three tiny incisions, the right tools, and the right procedures, and he was back on his feet just hours after an operation that used to send a patient into weeks of recovery. Home-remodeling projects can be similar. Careful planning can keep cutting and demolition to a minimum with little disruption to the living environment and the lives of the clients. And the job takes less time.

One such project is putting a new window into an existing home (see the photo at right). I try to perform most of the work from the outside, leaving the wallboard intact as an interior barrier against the mess, right up until the window goes in.

Upgrade with minimal mess. A new window improved the look and let in light. Working from the outside controlled the mess.

1. Strip Off the Siding

You'll be happy if the house has vinyl siding: It comes off easily. Available at most hardware stores, a zipper slips between siding panels to separate them.

After lifting the upper panel, a flat bar pops the nails of the siding panel below.

Siding in the window area comes off first. With this tool, vinyl siding is one of the easiest types to remove and replace.

Peel Back the Skin

The first step is to remove the siding in the area where the window is going to be located (see the photos above). If another window is on the same wall, I use it to gauge the head height for the new window. I also check around the corner to make sure that the new window will be at the same elevation as any existing window that's in the same room (because you never know).

I remove the siding, starting with the course above the window location and moving down to a course or two below the window. Because this particular house had vinyl siding, I knew I'd be removing full lengths of siding and exposing a large horizontal area, so I didn't have to be to fussy about the horizontal location at this point. If I had been working with sidewall shingles or with clapboards, I would have needed to be more precise about locating the window and

With all the siding removed, the housewrap is peeled back to reveal the wall sheathing underneath.

removing just a small area of siding around the spot of the new window.

To unlock the top row of vinyl, I use a tool called a zipper, basically a flat hook that is inserted behind the interlocking edges of two adjacent pieces. The tool grabs the lip of the top piece of vinyl and bends it out, unlocking it from the lower piece. Once the work has started, pulling the tool along the length of the siding separates the two pieces, opening them up like a zipper.

With the top piece of vinyl lifted up, I pull out the nails from the lower one. When all the nails are removed, I grasp the length of siding and push down, unsnapping it from the next course. The remainder of the siding is removed easily in the same fashion. Next, I carefully cut the housewrap and tack it out of the way until the opening is framed.

2. Lay Out the New Window Opening

Use existing framing if possible for part of the rough opening.

With the sheathing exposed, the sides of the rough opening are laid out first. In this case, an existing stud forms the left side of the opening, and the author measures over for the right side.

The top of the opening matches a window on the same wall.

Mark the Opening on the Sheathing

With the wall sheathing exposed, I now can locate and mark the rough opening precisely (see the left and center photos above). The existing stud locations are easy to determine by the nail patterns, and on this job the rough opening fell about 2 in. away from the edge of a stud.

To make life a little easier, I try to use existing studs for the sides of the rough opening, as long as the studs are close to plumb. (If they are way out of plumb, I adjust the opening to miss them completely and insert new plumb studs for both sides.) To land the opening on a stud, I had permission from the owner to move the window one way or the other, up to 6 in. In this case, I moved the opening about 2 in. until it fell along the edge of a stud. Using that edge as a start-

The bottom is measured down from the top point.

A circular saw with the blade set at a shallow depth cuts the sheathing.

ing point, I drew the rest of the rough opening on the sheathing.

Cut the Sheathing First, Then the Studs

Because this wall happened to be a gable end, I didn't have to install a load-bearing header and jack studs, which meant that I didn't have to take out a large section of sheathing. Using a circular saw with the blade set slightly deeper than the thickness of the sheathing, I cut and remove the sheathing from the opening. I set the blade shallow to avoid hitting anything that may be buried in the wall.

With the sheathing removed, I cut the exposed insulation just below the sill height and remove it from the bays, except the bay where I'm going to add a full-height stud. For that bay, I remove the full batt of insulation, cut it to be about 1 in. shy of the verti-

3. Prep the Opening from the Outside

Leave room for the sills when you remove the studs.

A reciprocating saw plunged through the sheathing 1½ in. below the opening cuts the stud at the right height for the sill.

With the old studs removed, screws attach the new framing to minimize wall damage inside.

New header and sill pieces go into the recess left by the reciprocating saw.

cal edge of the sheathing, and replace it in the wall cavity.

If the header and sill are just 2x4s on flat, I mark the sheathing 1½ in. up from the top of the opening and 1½ in. down from the bottom at each stud location (see the photos on p. 185). With a reciprocating saw, I plunge-cut through the sheathing and then through the full depth of the studs at my marks. I don't worry if the blade pokes through the drywall a little. The interior treatment should cover any place the blade might pierce through. Once the studs are cut, I knock the pieces out of the opening carefully with a hammer.

New Framing Slips behind the Sheathing

Because I used an existing stud for the left side of the opening, I had to insert a new stud only for the right side. For this stud I cut a length of 2x4 to go from the sole plate to the top plate.

I slide the new stud into the wall cavity, and then, starting at the bottom, I tap the stud gently into position. I fasten the studs in place by driving deck screws through the sheathing as well as through the stud and into the plates, if I can reach. Driving screws causes less vibration than nailing and minimizes the chance for drywall cracks inside. With the stud fastened securely, I replace any pieces of sheathing that had been removed for sliding the stud into the wall.

The next step is cutting and replacing the insulation in the newly formed stud bays above the 2x4 header and below the sill. After that, I measure, cut, and install the header and sill. As with the new stud, I screw the header and sill to the sheathing and toe-screw them into the support studs.

With the framing complete, I can take out the wallboard that has been protecting the inside of the house (see the photos at right). To mark the wallboard for cutting, I drive a nail through each corner of the framed opening from the outside. Inside the house, I draw lines connecting the nail holes to outline the opening.

To keep the gypsum dust to a minimum, I use my cordless trim saw with an old carbide blade to cut right through the wallboard with little effort. Its relatively slow blade speed creates no more gypsum dust than cutting by hand. After I cut and remove the wallboard, I screw the perimeter of the opening to the new framing.

Installing the Window and Buttoning Up

With the rough opening complete inside and out, my attention turns to the window

4. Wallboard Comes Out

Drywall on the inside protects against construction mess on the outside.

Inside, the corners are connected and the opening is drawn.

Nails driven into the corners from the outside transfer the opening to the inside.

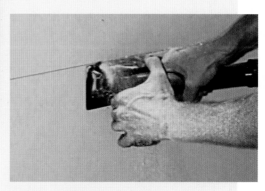

A trim saw with a carbide blade cuts the wallboard with minimal mess.

The slow blade speed of the Makita 9.6v circular saw (800-462-5482) lets it cut drywall without making a lot of dust.

The wallboard comes out, making way for the new window.

5. Install the Window and Wrap Up

If you've done your work carefully,
putting in the window is the easiest part.

Installing the window is an anticlimax after all the careful prep work. The window height again is measured down from the siding course to match the neighboring window.

itself. The installation and flashing are the same as with a normal window. For this window, I cut and bent aluminum flashing for the sill and used flashing membrane to seal the corners.

Making sure the head of the window lines up with its neighbor down the wall is easier with a helper (see the left photo above). The crew member inside simply shims the window to the right height, while the outside crew member measures down from the siding course above.

After the window is nailed in place, one crew member tackles the inside finish around the window. The crew member outside completes the flashing around the window and then reinstalls the siding. When the last row of siding has been nailed

Flashing and housewrap are woven back around the window and under the siding.

The siding is reinstalled and zipped back in place.

on, the zipper tool is used in reverse to mate together the rows of siding, and after four hours' work, the window looks as if it had always been there.

Rick Arnold, a contributing editor to Fine Homebuilding, *is a contractor and residential consultant in North Kingstown, Rhode Island, and the author of* Working with Concrete *(The Taunton Press, 2003).*

Hanging French Doors

■ BY GARY STRIEGLER

The old adage "Children should be seen and not heard" must have been coined by a French-door salesman. Most people I know like French doors. They're an elegant way to cordon off rooms without visually separating them. In fact, they act more like windows between rooms than doors.

Hanging French doors, however, can be frustrating. Getting two doors, either of which may be slightly warped, to meet up perfectly when installed in the almost certainly imperfect framing of a house requires several levels of adjustment. Correcting the rough opening or fixing an out-of-plumb, or cross-legged, opening might call for the use of a sledgehammer, which doesn't take much technical skill. Making adjustments for an out-of-level floor or a bowed door, on the other hand, takes a little more finesse.

Gary Striegler is the principal of Striegler and Associates, a custom-home contractor in Fayetteville, Arkansas.

Two doors are harder than one. The trick is setting one jamb leg plumb and just tacking the other before hanging the doors and finally adjusting the second side perfectly.

Assessing the Rough Opening

Rough openings are rarely perfect, but a door jamb must be plumb and level. Some properly placed shims will correct most rough openings.

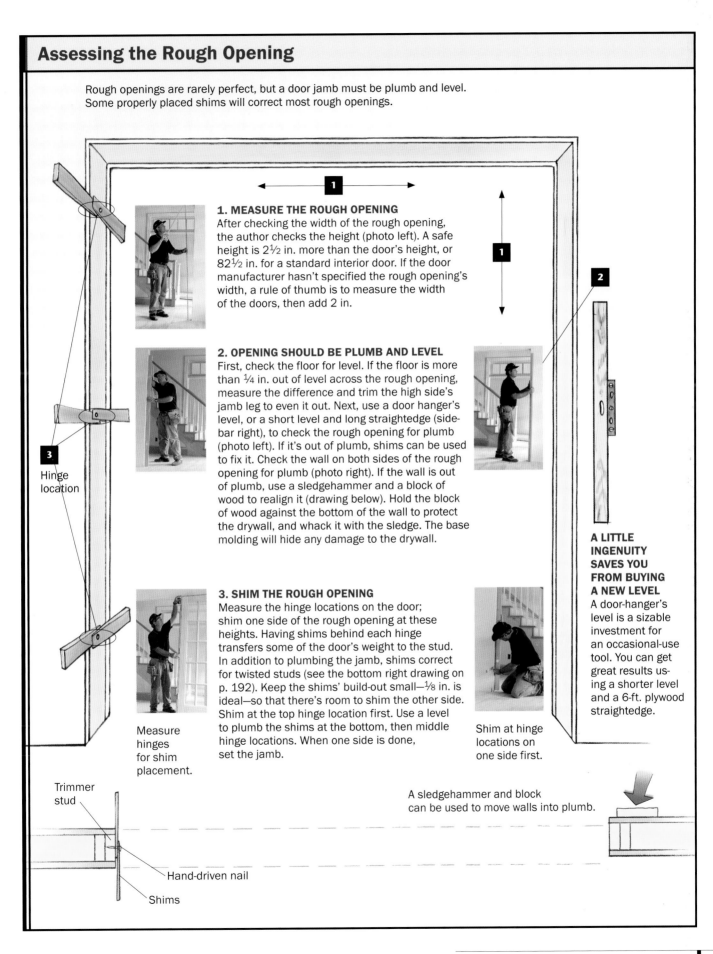

1. MEASURE THE ROUGH OPENING
After checking the width of the rough opening, the author checks the height (photo left). A safe height is 2½ in. more than the door's height, or 82½ in. for a standard interior door. If the door manufacturer hasn't specified the rough opening's width, a rule of thumb is to measure the width of the doors, then add 2 in.

2. OPENING SHOULD BE PLUMB AND LEVEL
First, check the floor for level. If the floor is more than ¼ in. out of level across the rough opening, measure the difference and trim the high side's jamb leg to even it out. Next, use a door hanger's level, or a short level and long straightedge (side-bar right), to check the rough opening for plumb (photo left). If it's out of plumb, shims can be used to fix it. Check the wall on both sides of the rough opening for plumb (photo right). If the wall is out of plumb, use a sledgehammer and a block of wood to realign it (drawing below). Hold the block of wood against the bottom of the wall to protect the drywall, and whack it with the sledge. The base molding will hide any damage to the drywall.

3. SHIM THE ROUGH OPENING
Measure the hinge locations on the door; shim one side of the rough opening at these heights. Having shims behind each hinge transfers some of the door's weight to the stud. In addition to plumbing the jamb, shims correct for twisted studs (see the bottom right drawing on p. 192). Keep the shims' build-out small—⅛ in. is ideal—so that there's room to shim the other side. Shim at the top hinge location first. Use a level to plumb the shims at the bottom, then middle hinge locations. When one side is done, set the jamb.

3 Hinge location

Measure hinges for shim placement.

Shim at hinge locations on one side first.

A LITTLE INGENUITY SAVES YOU FROM BUYING A NEW LEVEL
A door-hanger's level is a sizable investment for an occasional-use tool. You can get great results using a shorter level and a 6-ft. plywood straightedge.

Trimmer stud

Hand-driven nail

Shims

A sledgehammer and block can be used to move walls into plumb.

With one side of the rough opening plumbed, set the jamb. Additional shimming will be necessary to fine-tune the jamb for a perfect fit.

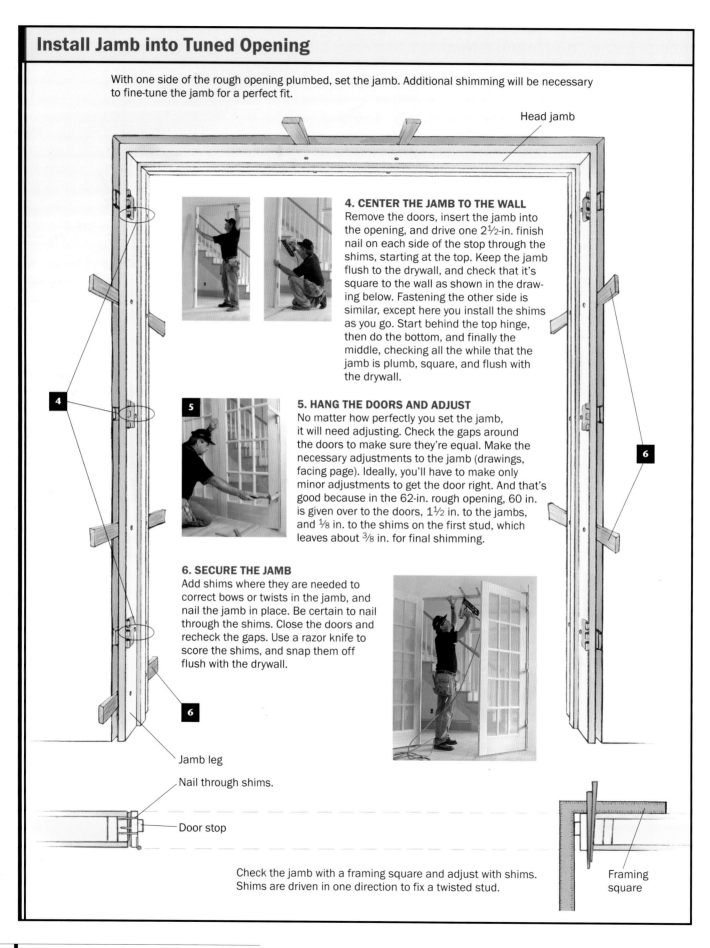

Head jamb

4. CENTER THE JAMB TO THE WALL

Remove the doors, insert the jamb into the opening, and drive one 2½-in. finish nail on each side of the stop through the shims, starting at the top. Keep the jamb flush to the drywall, and check that it's square to the wall as shown in the drawing below. Fastening the other side is similar, except here you install the shims as you go. Start behind the top hinge, then do the bottom, and finally the middle, checking all the while that the jamb is plumb, square, and flush with the drywall.

5. HANG THE DOORS AND ADJUST

No matter how perfectly you set the jamb, it will need adjusting. Check the gaps around the doors to make sure they're equal. Make the necessary adjustments to the jamb (drawings, facing page). Ideally, you'll have to make only minor adjustments to get the door right. And that's good because in the 62-in. rough opening, 60 in. is given over to the doors, 1½ in. to the jambs, and ⅛ in. to the shims on the first stud, which leaves about ⅜ in. for final shimming.

6. SECURE THE JAMB

Add shims where they are needed to correct bows or twists in the jamb, and nail the jamb in place. Be certain to nail through the shims. Close the doors and recheck the gaps. Use a razor knife to score the shims, and snap them off flush with the drywall.

Jamb leg

Nail through shims.

Door stop

Check the jamb with a framing square and adjust with shims. Shims are driven in one direction to fix a twisted stud.

Framing square

The Ideal Door Installation

French doors should have a nickel-thick gap all the way around. Wood doors shrink and grow with humidity. This gap ensures clearance for the doors to open and close even when swollen with moisture. If you want to be really good, gauge the humidity when you're installing the doors. If they're already swollen, make the gap between them a bit tighter so that it doesn't open objectionably when the wood dries out. If the doors are dry when hung, allow a little extra room for growth. If gaps are uneven, the drawings below show how to get them right.

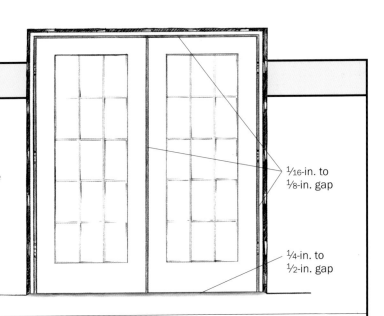

$\frac{1}{16}$-in. to $\frac{1}{8}$-in. gap

$\frac{1}{4}$-in. to $\frac{1}{2}$-in. gap

PROBLEM There's an uneven gap between the door and head jamb and wider gap at the bottom between the doors.

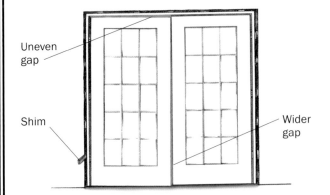

Uneven gap

Shim

Wider gap

SOLUTION Tap in an extra shim behind the bottom hinge as shown to even the gaps around the doors.

PROBLEM There's a wider gap between the doors near the head jamb and too little at the bottom.

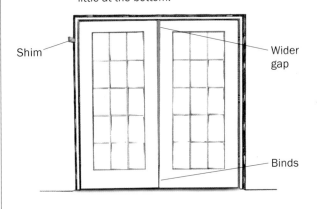

Shim

Wider gap

Binds

SOLUTION Shim behind top hinge as shown.

PROBLEM There's an even gap between the doors, but they aren't aligned.

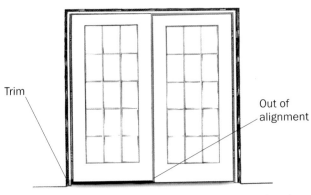

Trim

Out of alignment

SOLUTION One of the jamb legs is too long. With a handsaw, shorten the longer leg by the difference between the two doors. Then pry down on the top of the cut jamb leg. If it doesn't move, cut or pull the nails holding that leg, lower it, and renail.

PROBLEM The doors aren't in plane.

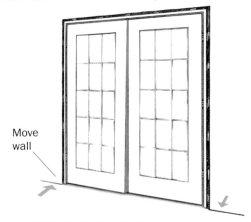

Move wall

SOLUTION Check the walls for plumb, and adjust with a sledge as in the bottom right drawing on p. 191. If both walls are plumb, a door is bowed. Replace it, or split the difference by knocking each wall a little out of plumb.

A New Door Fits an Old Jamb

■ BY GARY M. KATZ

I used to hate hanging doors. It's way too easy to make a mistake, like hinging the door backward or upside down, or planing the bevel in the wrong direction. And I always had a tough time making a ⅛-in.-to-zero cut across the top of a door, especially if the door cost more than all my tools combined and especially if someone was watching me work.

But years of hanging doors and learning from professional door hangers have tempered my views. In fact, the techniques that I outline here can make door hanging foolproof and fun, even if you're hanging your first door. Using just a few simple tools and following the steps in order, you'll no longer need to be afraid of doors—or of people watching you work on them.

Gary M. Katz is a carpenter living in Reseda, California, and the author of For Pros by Pros: Hanging and Installing Doors *(The Taunton Press, 2002).*

1. Shims Position the Door in the Opening

Scribing the door to the opening is an important first step. Start by setting the door on a couple of shims, then hooking it against the top of the jamb (see the photo on the facing page). A homemade door hook holds the door against the jamb (see the drawing at right).

Adjusting the shims raises or lowers the door to the right position. If the head of the jamb is out of level, raise or lower one side of the door until the top rail is parallel to the jamb head. A small prybar moves the door until it's centered in the opening (see the photo below), but I leave at least 3⅞ in. on the lock stile after planing; otherwise, some dead bolts might not fit. If the head is out of level and the jamb can't be fixed easily, cheat the door a little out of plumb to make the head look better.

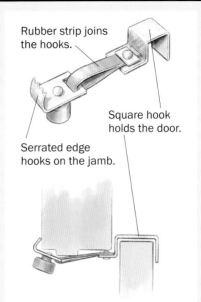

Rubber strip joins the hooks.

Square hook holds the door.

Serrated edge hooks on the jamb.

2. Mark the Door Edges

Use a simple set of dividers to scribe the edges of the door. Because I'm on the stop side (see the photo at right), I spread the dividers to $3/16$ in. (for a $1\frac{3}{4}$-in. door) to account for the hinge gap as well as bevels on both edges. If I'm fitting an interior door, I scribe the bottom of the door for the floor it swings over. For standard carpet, I spread the scribes to $1\frac{3}{8}$ in. An exterior door is scribed for the threshold and door shoe (see the photo and the drawing below).

Before taking down the door, make a large X out of tape on the hinge-side top of the door to orient it once you've carried it to the door bench. Also measure for each hinge location. I used to transfer hinge locations from the jamb to the door by eye, but because the trim often keeps the door $\frac{1}{2}$ in. away from the jamb, it's hard to keep the marks perfectly level. Careful measurements eliminate guesswork.

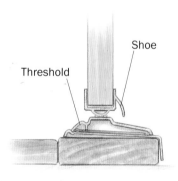

Threshold   Shoe

3. Trim the Top and Bottom First

A door bench is helpful when working on a door, but sawhorses also can be used. After placing the door flat with the X facing up, score the cutline across the stiles with a razor knife to prevent tearout. Cut about 1/16 in. wide of the line, using a sharp blade in your circular saw (see the photo above). Then plane right to the line. I start with the plane upside down, stopping within a foot of the opposite end so that I don't blow out the back edge of the stiles (see the photo at right). After planing to the line, turn the plane right side up and finish the other end with no blowout. While the door is lying flat on the bench, seal the top and bottom.

4. Bevel and Mortise the Hinge Edge

Set the door on edge with the X facing the bench to plane the hinge stile. Plane to the scribe line with the plane set to about a 3-degree bevel (see the photo below and the drawing on the facing page). The bevel prevents hinge binding. I also ease the edges with the plane at a 45-degree angle.

Lay out the hinges with a tape measure and look for the X to make sure you lay out the mortises from the top (see the top right photo on the facing page). The hinge barrels should point away from the X. For speed and accuracy, I use router templates for all hinge and hardware mortises (see the bottom left photo on the facing page). To keep the hinge screws from splitting the stile, drill pilot holes with a centering bit before attaching the hinges (see the bottom right photo on the facing page).

3-degree
bevel

Because I keep the scribe lines and the X facing the door bench at all times, I flip the door vertically, not horizontally, to work on the opposite edge, the lock stile (see the left photo above). Bevel the lock stile with the plane at the same 3-degree bevel (see the top right photo above). Then measure for the lock bore, again remembering to look for the X so that you measure from the top instead of the bottom. Because I install lots of doors, I've invested in a lock-boring jig, although a paddle bit and a hole saw also can work well. Drill and mortise for the lockset (see the bottom right photo above), then carry the door back to the opening. About 95 percent of the time, the door fits the first time I swing it.

6. Final Tweak for a Perfect Fit

To hang the door, tip it until the top screw hole in the top hinge lines up with the screw hole in the jamb, and drive that screw. Align the door by pushing with your foot, and drive the rest of the screws in the hinges (see the photo at left).

When the door is swinging from the jamb, check the fit. Doors often need tweaking because hinges aren't all the same. To move the door toward the strike jamb, place a nail set between the hinge leaves and against the hinge barrel (see the bottom left photo), then close the door gently. The nail set spreads the hinge slightly.

To move the door toward the hinge jamb, first pop up the hinge pin until it's just engaging the top barrel. Then tighten an adjustable open-end wrench around each of the barrels on the hinge leaf attached to the door (see the bottom right photo). Use the wrench to bend the hinge toward the strike side. This will close the gap on the hinge side.

Installing Bifold Doors

■ BY JIM BRITTON

1 Start by assembling the jamb. First, the author spreads out the jamb parts in front of the closet. Then he uses a pneumatic stapler to anchor the side jamb to the head jamb.

2 A 1x strip hides the door track. Once the jambs are assembled, the author attaches a 1x2 to the inside of the head jamb. The door track tucks against the backside of this strip.

3 Next, nail the casings to the jamb stock. Leaving a $3/16$-in. reveal around the inside edges of the jambs, the author readies the jamb assembly by nailing on the casings.

If you work as a trim carpenter long enough, sooner or later you'll come to the realization that prefabricating components on a bench makes it easier to put them on the wall. I put this approach to work in all my trim-carpentry tasks. But in the case of installing bifold doors—the ubiquitous accordion-style panels that conceal many an American closet—the workbench is the floor in front of the closet. The job starts with the jambs.

Unlike prehung doors, which are hinged to their jambs when they leave the factory, bifold doors are typically shipped without the jambs. Instead, bifold doors are hinged

to one another and packed in a cardboard box. The box also contains a bag full of hardware for hanging the doors. The doors illustrated here, for example, included the 2400 series Bi-Fold Door Hardware from Stanley®. Detailed instructions for adjusting the hardware accompany the doors.

I typically purchase bifold doors from the same shop that supplies my passage doors. Along with the doors, I have the shop send along finger-jointed pine jamb kits for the doors. I prefer pine because nails driven into pine—unlike medium-density fiber-board (MDF), the other jamb-material option—don't leave telltale dimples that need sanding.

4

Shims elevate the jambs. With a spirit level long enough to span the rough opening, the author assesses the floor for level and places shims accordingly next to the trimmer studs. The shims, which are at least ³/₈ in. thick, lift the jambs enough to accommodate the carpet.

5

Lift the pretrimmed jamb into its opening. With the bottoms of the side jambs sitting on their shims to ensure a level head jamb, the author slides the jamb assembly into the rough opening. Affixing it to the rough-opening frame starts from the top down.

Center the head jamb. Next, the author slides the assembly from right to left, and marks the positions with a pencil. Splitting the difference centers the jamb. One nail through the casing into the header maintains alignment at this stage.

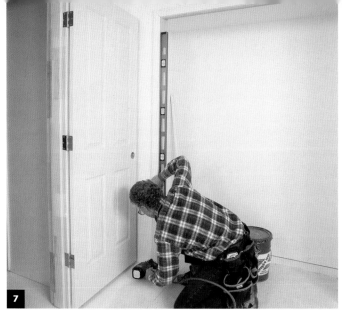

Plumb the side jamb. The author uses the 6-ft. level to check the side jamb for plumb, moving the bottom of the jamb toward or away from the trimmer until the bubble is centered. Once he's got it right, he affixes the casing to the trimmer with a nail near the bottom of the jamb.

Shim the jamb, and nail it home. Once the jamb has been plumbed, it can be permanently affixed to the trimmer. The author uses a 15-ga. nailer, driving the fasteners through shims to keep the jamb from deflecting toward the framing as the nails slam home.

Now go to the other side. The bottom of the opposite jamb is located with a tape measure. The distance between the jambs at the top should be repeated here. Double-check for square by measuring the diagonals from casing corners to jamb bottoms.

Rough openings should be sized in a manner similar to those for passage doors by adding 2 in. to the nominal dimension of the doors. For example, a 5-ft. 0-in. bifold requires a 62-in.-wide rough opening. That leaves enough room for the jambs (two at ⅝ in. each) plus some wiggle room for plumbing the jambs. I frame my rough openings to be 82½ in. high from the subfloor. That's a standard 81-in. trimmer atop a 2x plate. Sharp-eyed carpenters will note that even though they are called 6-ft. 8-in. doors, bifold doors are really 6 ft. 7 in. tall. The missing inch allows room for the track at the top of the doors.

10

The head casing gets nailed on last. The author takes a slight bow out of the head casing by flattening it against the level as he nails the casing to the header.

11

Screw the track to the head jamb. The door track fits flush against the 1x2 nailed to the inside of the head jamb. If the screws reach the header, be sure to shim the jamb to keep it from deflecting.

12

Align the jamb brackets with the door track. At the base of each side jamb, an L-shaped bracket supports a pivot pin mounted in the bottom of the pivot door. The serrations in the slot capture the star-shaped pivot pin at the bottom of the door, holding it at the desired distance from the jamb. The bracket sits atop a ³⁄₈-in.-thick plywood block, which keeps the bracket from being buried by the carpeting. A pencil mark on the side jamb notes the centerline of the door track above.

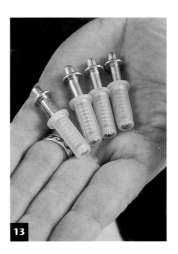

13

14

The doors swing on pins. At the top, each door has a spring-loaded pivot pin (top) that slides in the door track. At the bottom, the door closest to the jamb has an adjustable pin that fits into the jamb bracket (bottom).

The jamb kits often include 1x1 moldings that are meant to conceal the edges of the doors around the sides and the door track. I think they're ugly. These clunky 1x1s give a tacked-on look to the trim that isn't in keeping with the look of the passage doors. So I leave out the 1x1s, preferring instead a detail that I learned from an ace finish carpenter back in the 1970s. As shown in photos 2 and 11, I add a 1x2 to the head jamb, and then conceal most of it with the same trim that cases the passage doors. I don't use any stops along the sides of the doors because the doors can easily be adjusted to maintain an even gap between their stiles and the jambs when the side jambs are aligned correctly.

Install the doors from the top. Insert the top pivot pins into their slots, and push up to compress the springs. This makes room for the lower pivot to clear the jamb bracket as its pin is located in the serrated slot.

Bifold doors swing on pins instead of on the leaf hinges common to passage doors. The pins fit into predrilled holes in the top and bottom of the doors. With this system, most of the loads exerted by the doors are delivered to the pivot brackets at the bottoms of the jambs. So don't worry about heavy structural nailing through the jambs to keep the doors from sagging. You'll note in the photos that I put blocks under the pivot brackets to keep them from being buried by carpeting. If the floor is to be hardwood, leave out the blocks and install the doors after the hardwood is in place.

Hardware varies a bit from brand to brand, but the principles are the same. The

16

17

Ready for knobs. The door closest to the jamb in a pair of bifold doors is called the pivot door. Its mate is called a guide door. You'll get the best leverage for opening and closing the doors by affixing the knobs to the pivot doors.

Aligners keep the guide doors flush with one another. Metal tabs called aligners are installed about 12 in. above the floor on the backside of the bifold doors. As the doors are brought together, the aligners engage one another to snap the doors closed.

door closest to the jamb is the pivot door. Its mate is the guide door. The pivot bracket at the bottom of the jamb is fixed, but the pin can be moved along its serrated slot to fine-tune the pivot door's distance from the jamb. The pin at the top of the pivot door fits into a bracket in the door track that can be adjusted with a set screw. The pin in the

guide door fits into a sliding, spring-loaded guide that snugs against the other pair of doors, holding them tightly together when closed.

Jim Britton is a carpenter and general contractor in Jacksonville, Oregon.

Installing Prehung Doors

■ BY JIM BRITTON

Of all the tasks a trim carpenter faces, few offer the opportunity to transform the look of a house quickly from ragged edges to finished surfaces like installing prehung doors. It's the trim carpenter's version of instant gratification because once in the groove, a good trim carpenter can install a door, its jamb, and all the casings in about 15 minutes. That's money in the bank for a pro, and a satisfying slice of sweat equity for the owner/builder.

But doors that squeak, bind, stay open, or swing open by themselves are constant reminders of the fallibility of the trim carpenter. Here, I'll describe the methods I've settled on after 20 years in the trades for efficiently installing a typical prehung door and avoiding common glitches that bedevil a door installation. Like most home-building jobs, installing a door begins with checking work done before you got there.

Check the Rough Opening First

In a perfect world of accurate levels, conscientious framing crews, and straight lumber, all rough openings are square, plumb, and correctly sized. Because these three conditions rarely coincide, it falls to the trim carpenter to compensate for less-than-perfect rough openings.

Although there are exceptions, the rough opening should be 2 in. wider and 2½ in. taller than the door. Thus the correct rough opening for a 2-ft. 6-in. door would be 32 in. by 82½ in. The extra space allows room for the door jambs and a little wiggle room to accommodate rough openings that are out of plumb. In my experience, rough openings are the same for both interior and exterior doors that are made by door manufacturers. Doors made by window manufacturers, on the other hand, sometimes require a different rough opening. If in doubt, check with the manufacturer before the framers start work.

Before installing a door, I inspect the rough opening to familiarize myself with its condition. First I check the dimensions to see if they are workable. Then I use a 6-ft. level to check that the two trimmers (the studs that frame the rough opening) are plumb in both directions (see photos 1 and 2). Sometimes the trimmers will actually be plumb in both directions, in which case the door jambs will be flush to the wall, and the casings will be easy to install.

But in some situations, the wall will be out of plumb in section, with the trimmers plumb in elevation. In this case, the door jambs will have to protrude slightly beyond the plane of the wall at the top and bottom on opposite sides.

Another common condition is the parallelogram-shaped rough opening. The wall may be plumb in section, but the elevation view of the rough opening is out of plumb. The net door width is usually ½ in. narrower than the rough opening. Therefore, I can install a plumb door in a rough opening that is up to ½ in. out of plumb. The jambs will fit snugly to the diagonally opposite corners of the rough opening. If the rough opening is more than ½ in. out of plumb, I use a sledgehammer to pound the trimmers into line. I can usually move a trimmer up to ½ in. without

1 and 2. Read the rough opening. Before installing the door and its jamb, the author checks the trimmers on both sides of the rough opening with a 6-ft. level to see if the trimmers are plumb. If the jamb needs to project beyond the plane of the wall in order for the door to hang plumb, he notes the direction of the adjustment on the trimmer. To avoid mistakes, he marks on the floor the direction the door will swing.

3. Check the trimmers for twist. If the trimmer isn't square to the header, the door jamb will also be askew. Use a square to gauge the accuracy of the door frame.

adversely affecting the drywall. This operation requires cutting back the sole plate once the trimmer has been adjusted.

The scissor condition, in which the trimmers are out of plumb in opposite directions in section, requires a more involved solution. Let's say one trimmer is ½ in. out top to bottom in one direction, and the other trimmer is out ½ in. in the other direction. This situation amounts to 1 in. of scissor. This condition is remedied by holding the jamb out ¼ in. at the top and in ¼ in. at the bottom on opposite sides of the wall. Do the opposite for the other trimmer.

The other condition I look for is twist (see photo 3 on p. 209). If the trimmers aren't square to the header, the jamb will likewise be twisted. This condition results in hinge binding or a poor visual relationship between the door and its jamb after the installation. I take the twist out when I affix shims to the trimmer.

Next, Put Up the Hinge Shims

To begin an installation, I measure the height of the top and bottom hinges of the door from the bottom of the hinge jamb. My 6-ft. level makes a convenient stick to note the middle of the hinge positions (see photo 4). These locations mark where I fasten my shims to the framing before the door goes in.

I shim the bottom location first with an appropriate combination of shims to bring the level plumb and to compensate for any twist in the trimmer. Then I move to the top hinge position, holding the shims in place with my level as I affix the shims to the trimmer (see photo 5). Prenailed shimming makes handling the door easy and ensures that the door will automatically be plumb in the elevation view. I use a 15-ga. or 16-ga. pneumatic nailer loaded with 1¾-in. to 2-in. nails for installing prehung doors. If you don't have one of these wonderful time-saving tools, use 8d finish nails instead.

Now it's time to squeeze the door and jamb into the opening. Remove any nails or straps used as bracing, and place the hinge jamb atop the thick end of a shim resting on the floor next to the trimmer (see photo 6). Raising the jamb has three benefits: It eliminates squeaks by separating floor and jamb, it eliminates the problem of an out-of-level floor preventing the strike jamb from not coming down far enough to engage the lockset latch, and it eliminates (or minimizes) the need to remove some of the door's bottom to accommodate finish flooring.

Once the jamb is in the rough opening, I swing open the door. If it's a troublesome installation, I'll block the door with a couple of shims. But typically I leave the open door unsupported. If the wall is plumb at the rough opening, I bring the edge of the hinge jamb flush with the drywall and nail it to the trimmer through the shims. If the wall isn't plumb, I compensate for the error by moving the jamb out equal amounts at the bottom and then the opposite direction at the top. I make pencil marks on the shims to note the correct alignment for the edge of the jamb (see photo 7).

I affix the jamb in the correct position relative to the wall with a couple of nails through the top and bottom shims. Then, while the door is still open, I drive a couple of nails through the jamb right next to the hinges (see photo 8). The hinge jamb and door now should be hanging plumb because they are held fast against the shims. If there is a middle hinge, shim it at this time, taking care not to make any changes to the already perfect alignment.

Nails are enough to keep the jamb of a hollow-core door from pulling away from the trimmer. But if I'm hanging a solid-core door, I run a 2½-in. screw through the jamb and into the trimmer next to each hinge. Because this step leaves a hole that no painter will be pleased to discover, I put the screws under the doorstops. Working from the bottom, I carefully pry away the stops and set them aside (see photo 9).

CHECK THE ROUGH OPENING FIRST

4. Note the hinge positions. Using a 6-ft. level as a story pole, the author marks the centers of the top and bottom hinges.

5. Affix the shims. Shims behind the top and bottom hinges make backing for the jamb. A single shim at the top compensates for twist.

6. Don't forget the shim on the floor. Elevate the hinge jamb by placing it atop the butt end of a shim shingle as the door is lifted into position.

7. Mark the jamb alignment on the shims. If you need to adjust the edge of the jamb in or out of the plane of the wall to get the door to hang plumb, make a note of the correct position of the jamb's edge on the top and bottom shims.

8. Nail the jamb. Secure the jamb to the trimmer with a couple of nails right next to the hinges. The nails must pass through the shims.

9. Remove the stops. Pry the doorstops from the bottom up. Then locate the screws that secure the jamb to the trimmer under the stop.

I sometimes run a 2½-in. screw through a hinge and into the trimmer. But I don't do this to keep the door from sagging: A properly hung door doesn't sag. Instead, I use the longer screw to straighten a warped jamb or to compensate for a hinge mortise that might be shallow. The longer screw will give me about ¹⁄₁₆ in. of adjustment.

10. Equalize the strike-jamb reveal. Use shims placed next to the door latch to adjust the strike jamb in or out until the gap is consistent from top to bottom.

11. Reinforce the strike plate. After pulling the doorstop, run a screw through the strike jamb next to the door latch.

10

11

Secure the Head Jamb and Strike Jamb

Now that the hinge jamb is firmly secured to its trimmer, I close the door. Next, I set the head jamb parallel to the top of the door by raising or lowering the strike jamb. At this stage of the game, a single, unshimmed nail through the strike jamb into its trimmer or through the head jamb into the header can help hold the parts in alignment while I assemble the correct combination of shims. A jamb held by a single nail still can be pried in or out as needed. A shim under the strike jamb also can be helpful.

When I've got the head jamb parallel with the top of the door, I check the reveal along the edge of the door and the strike jamb. I put a couple of shims between the jamb and the trimmer 6 in. down from the head jamb and adjust the shims until the gap, or reveal, between the door and the jamb is the same at the top corner. Then I shim the bottom of the jamb, 6 in. from the floor, and the center of the jamb opposite the strike plate (see photo 10). Some door jambs are straighter than others. If I've got one with some dips and wows in it, I add shims as necessary to keep the reveal consistent. I add extra support to the jamb where the strike engages it. To do so, I pry away the doorstop and drive a 2½-in. screw into the trimmer (see photo 11).

Replace the Doorstop

The door is now where I want it and fully supported. With the door closed flush with the head jamb, I position the head stop on the hinge-jamb side with the help of a dime (see photo 12). This ¹⁄₁₆-in. gap between the door and the stop helps keep the door from binding on the stop and allows for paint buildup. I continue this space down the hinge jamb with the hinge stop, attaching the stop with my 16-ga. nails, 16 in. on center.

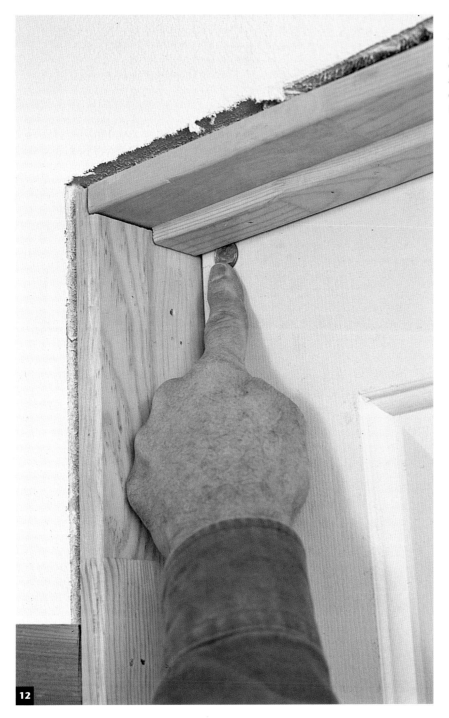

REPLACE THE DOORSTOP
12. Don't forget the stop gap.
Use a dime between the closed door and the doorstop to gauge a consistent gap between the door and the stop. The gap allows for paint buildup.

I install the strike stop so that it just touches the entire height of the door with the door's face and the jamb flush over the full height. This system works well for a strike that has an adjusting tang. However, if the strike will be the T-mortise type, it will have to be installed first and the stop set to it.

Apply the Casings Last

I start casing a door at the top with the head casing set back $3/16$ in. from the edge of the jamb (see photo 13 on p. 214). The casing has 45-degree miters on each end, and the short side of the casing is $3/8$ in. longer than

13

14

APPLY THE CASINGS LAST

13. Casing starts at the top. The author begins trimming a door by installing the head casing first. He affixes the casing with pairs of nails on 16-in. centers. One nail goes into the jamb, and another into the header.

14. This gap won't make the cut. When the door jamb and the wall are in slightly different planes, the casings don't lie completely flat. The tapered gap at the inside corner is the result.

15. Undercut the side casing. With the side casing face-side up and slightly tilted, the author removes material from the miter cut with a bench-mounted disc sander.

15

the jamb opening. That gives me the ³⁄₁₆-in. reveal along both of the side jambs.

The tricky part of casing a door is dealing with the differential between the plane of the wall and the plane of the jambs when you've made allowances for an out-of-plumb rough opening. For example, this door had a head casing whose edge was recessed a bit from the plane of the wall. When I test-fit the side casing, I came up with a gap at the inside corner (see photo 14). To fix it, I undercut the miter with a disc sander (see photo 15). This cut isn't a back bevel, however. In this case I removed material from

the casing's face. Once I'd shaved the miter, I had an acceptable joint for paint-grade trim work (see photo 16). To keep the adjoining casings in the same plane at the outside corner, I put a thin shim under them (see photos 17 and 18).

I attach the side casings with pairs of nails, one into the jamb and one into the trimmer a couple of inches away. This nailing pattern helps ensure that the casing will lie flat. I nail the casings next to the hinges and the door strike because these spots are well-backed by shims. Nailing the casing at these points also reinforces the jamb.

16. Now it fits better. By undercutting the side casing with the disc sander, the author achieves an acceptable miter. A dose of caulk will touch up the remaining crevices.

17. Shim problem casings at the corners. If the casings are out of plane, slip a shim under the corner so that both pieces bear on it. Then trim the shim flush with the casing with a utility knife.

18. Nail 'em. Once the shim has been trimmed, secure the casing corners to the wall with nails driven into the header, trimmer, and door jamb. Fill any gaps between the wall and the casing with caulk.

As you can imagine, drywall edges can be a pain in the neck when the door jamb is below the plane of the wall. The hollow milled into the back of the casing is there to compensate for this situation. If the hollow isn't enough to accommodate the drywall, I use my hammer to "tenderize" protruding drywall edges.

If the floor is to be covered with carpet, I hold the side casings ⅜ in. above the floor. That gap gives the carpet guy some room to tuck the edges of the rug. It's a good idea to put a shim between the jamb and the trimmer at the bottom of the jamb if the room

is to be carpeted. The shim keeps the jamb from being deflected by the carpet layer's bump hammer as he tightens the carpet against the tack strips.

If the floor is going to be finished with ¾-in. hardwood strips, I set the side casings on ¾-in. blocks. When the floor is installed, the blocks come out, and the flooring slips into the gap.

Jim Britton is a trim carpenter and a contractor living in Fairfield, California.

Installing
Sliding Doors

■ BY GARY M. KATZ

I recently got a call from a homeowner who wanted me to come out and adjust a sliding door that wasn't working right. It sounded like a small job, so I figured that my toolbox and cordless drill were all the tools I'd need. I was wrong.

Right away, I had trouble unlocking the sliding door. The homeowner explained that the only way she could get the door to latch was by slamming it as hard as she could against the jamb. With the unlocked door close to the jamb, I could see part of the problem. The door was touching the jamb at the top but was more than ¼ in. away from the jamb at the bottom. I said to myself, Hey, this'll be a snap. I'll just raise the front wheel on the door, and the lady will think I'm a genius.

But when I slid the door back to get at the wheel-adjustment screw, the back of the door was rubbing so badly on the head jamb that the slider would open only halfway. I said to myself, Hey, this'll be a snap. I'll just lower the back wheel, and the lady will think I'm a genius. But the rear wheels were already set as low as they could go. That's when I noticed that the head jamb was pushed way up in the center.

Then I got down on my hands and knees, sighted down the sill and saw the real problem. The oak sill looked like a foothill in the San Gabriel Mountains. An improperly set sill is a classic, common mistake made while installing prefit sliding-door units. By following just a few a simple steps, you'll be able to avoid that mistake, and a few others, too.

Start at the bottom. The first step in installing a slider is leveling the threshold. Lay a long level on the rough opening, and shim the threshold up to the proper level.

Make Sure the Door Is Going to Fit

I admit it: I've been embarrassed by removing an old door before realizing that the new unit wasn't going to fit. So now I always measure the opening or, on a remodel, I measure what will become the opening before the door is ordered.

I like to have the rough opening about ½ in. larger in width and height, though some slider manufacturers call for rough openings a little bigger, in some cases 1 in. wider and taller than the unit dimensions. Sometimes the extra space is more than necessary, and sometimes it's not nearly enough if the floor of the opening is out of level or has a big hump in it.

If I'm working on a concrete slab that is grossly out of level or if the threshold has a terrible high spot, I know I'm going to need more head clearance. I look for these problems before assembling the frame by checking the threshold with a long level. If there is an old door in the opening, I check the floor just inside the threshold for obvious humps or for an out-of-level floor. It's also a good idea to give the jambs a quick check. If they're out of plumb, I may need to make the rough opening a little wider.

Although the wood sills on most sliders are back-primed by the manufacturer, some still need to be back-primed or sealed by the installer. If the sill isn't sealed properly, it will warp and twist. If the sill hasn't been thoroughly primed, I start back-priming as soon as I'm sure that the slider is the right size and that it won't be going back to the store. If possible, I start back-priming before removing the old unit so that I can get at least two good coats of exterior primer on the sill before it's time to assemble the frame. If the whole frame is wood, I also make sure that the bottoms of the side jambs are back-primed to keep the end grain from soaking up moisture.

Level the Sill in the Rough Opening

I begin prepping the opening by cleaning the threshold. Then I lay down the longest level I have that will fit between the jambs (see the photo on pp. 216–217). A short level placed on top of a straight board also works well. I shim up the low end until it's perfectly level; then I fill in the gaps between the level and the floor with additional shims placed about every 12 in. On concrete slabs I hold each shim in place with a blob

A paper drain pan. Special waterproof-paper flashing is layered in beds of silicone to form a drain pan under the slider.

of silicone adhesive. On wood floors I nail the shims down so that they won't move when I slide the frame in and out during the dry fit. I always double-check the header height from the top of the shims. If there's not enough room, I'll have to get a shorter slider. Next, I make sure the trimmer studs that flank the opening are plumb. I measure the width of the opening, and I add furring to the trimmers if the opening is too wide.

The next step is flashing the threshold to keep water from coming in under the door. If I've had an aluminum drain pan made for me, I press it into a bed of silicone right over the shims, making sure that corners and seams are well sealed.

If no aluminum drain pan is available (which is the case on virtually every remodeling job where I install a slider), I make the pan out of layers of Moistop® (Fortifiber Corp., 1001 Tahoe Blvd., Incline Village, NV 89451; 800-773-4777), a fiber-reinforced waterproof-paper flashing with a polyethylene coating on both sides. Moistop comes in 6-in.-wide rolls. I cut the first layer of paper flashing about 12 in. longer than the opening so that ends extend up the jambs, and I install that layer in a bed of silicone over the shims (see the photo on the facing page). I let a few inches of the paper flashing lap over the outside edge of the sill. Staples keep the flashing in place while the silicone adhesive is curing.

I cut a second layer the same length and bed it in silicone over the first layer, only this time I let the excess paper flashing extend into the room. If there is hardwood flooring or a subfloor that the sliding door will fit against, I wrap the excess flashing up the edge of the flooring to create a dam. If there is no flooring or subfloor, I leave the excess flap until after the door is installed, and then I staple it to the inside edge of the slider threshold to form the dam. Finally, I caulk a short piece of paper flashing into each corner, making sure that the stacked-up layers provide complete coverage.

Drill anchor holes during the dry fit. After the frame has been tested in the opening, anchor holes are drilled through holes in the sill with a masonry bit.

A Dry Fit Locates the Anchor Positions

The frames for most sliders—no matter whether they're wood, metal, or vinyl—have to be assembled on site. I lay all the pieces on a flat, open area with the outside facing up, and I screw the corners together after sealing them with silicone.

Before bedding the frame on top of the drain pan, I test the assembled frame in the opening to be absolutely sure that the door is going to fit right. If screws and concrete anchors are being used to secure the sill, a dry run is the best way to locate the anchors.

After centering the frame in the opening, I make sure that there's ample room to plumb the side jambs. If I'm installing the door over concrete, I mark both ends of the sill so that I can put it in the exact same position when I install it for good. For aluminum or vinyl sills, I run a masonry bit through the factory-drilled holes, drilling into the concrete for the plastic anchors (see the photo above). I locate concrete

Slider frame is bedded in caulk. Caulking is applied around the framed opening, and the nailing flanges on the side jambs are pressed into the caulk and fastened temporarily until the jambs are plumbed.

anchors the same way for sills on wood sliders, only first I counterbore the holes for wood plugs, which I install later to cover the screw heads.

After all the anchors are located, I remove the frame from the opening, sweep out the dust and dirt, and then insert the plastic concrete anchors. Before slipping the frame back into the opening, I lay down another bed of silicone caulking on top of the paper flashing. For sliders with a nailing flange or an extruded exterior trim that sits on top of the finished wall, I run a heavy bead of silicone caulking behind the flange or trim before installing the frame (see the photo above). I try to keep that bead of caulking back from the exposed edge of the trim so that the silicone doesn't squeeze out when I press the door into place.

A Transit and Fishing Line Make Straight Sills

Once the unit is in the opening and is sitting at the pencil marks I made during the dry fit, I tack each side jamb in place with a screw or with a half-driven nail through the face of the jamb or the nail flange. The slider sills that are 8 ft. long or less stay pretty straight, and I can usually secure the sill in place without adding shims.

For sills longer than 8 ft., I stretch a string between the jambs to make sure the sill is perfectly straight. The fluorescent nylon string that is common on construction sites is really too heavy for this task, sort of like using a framer's pencil for finish carpentry. My favorite string for straightening sills is

20-lb. braided Dacron® fishing line, the strong, thin backing that I use on my fly reels (available at any fishing store). I stretch the line tight between the side jambs and insert shims from the outside between the sill and the drain pan until the sill is even with the line.

Fortunately, almost all sliding doors have adjustable wheels. But for those that don't, the sills have to be set absolutely level and straight. The same is true for multiple adjoining units that have to be set accurately to ensure that the mullions and casings line up horizontally. For these situations I level both ends of each sill with a transit to put all the adjoining units at precisely the same level. Once the ends of the adjoining sills are shimmed to the same level, I stretch my string and insert any shims needed to make each sill perfectly straight.

After the sill is set, I work on the side jambs, shimming out each one until it touches the level evenly top to bottom (see the left photo below). I secure the side jambs with more screws or half-driven nails and wait until the door panels are installed before I fasten the side jambs permanently. After the side jambs have been shimmed, I measure the diagonal distance of the frame corner to corner as a final test to make sure the frame is square.

On sliders wider than 8 ft., I also test for cross-legged jambs, which happens when the jambs are not in precisely the same plane. With cross-legged jambs, the sill and the head jamb aren't parallel, and the doors panels might bind in their channels and won't slide smoothly no matter how much wax or silicone I use. I test for this condition by stretching strings diagonally between the corners (see the right photo below). I move the top of one jamb and the bottom of the opposite jamb in or out until the strings just touch in the middle.

For sliding doors without adjustable wheels, the sills have to be absolutely level and straight.

Shims bring the jamb out to the level. After the sill is leveled and the jambs are plumbed top to bottom, shims are inserted behind the jamb until the jamb touches the level.

Checking for cross-legged jambs. Strings stretched from corner to corner will touch in the middle if the jambs are in precisely the same plane, which ensures the door slides smoothly.

Most Door Panels Go in Top First

If the slider didn't come preassembled, the next step is putting in the door panels. The stationary panel is almost always installed before the active or sliding panel, and all stationary-door panels go into the frame top first.

In most cases, the head jamb has a channel that the panel fits into. After the door panel is lifted into the channel, the bottom is placed over the track on the sill. The panel is then slid into position against the jamb. Most sliding-door units have an additional threshold strip that snaps into the sill between the stationary panel and the strike jamb to lock the panel in place. The active panel then is installed top first in its own channel (see the photo below), and the wheels are adjusted down to lift the door off the track.

Adjust the Wheels to Make the Slider Parallel to the Jamb

The next step is adjusting the wheels on the active door. Some wheels have adjustment screws that are accessed through the leading and trailing edges of the door, and other types have adjustment screws accessible through the exterior face of the door just above each wheel. Turning the adjustment screws raises or lowers the wheels, which in turn affects the alignment of the door panel in the frame (see the photo on the facing page).

I adjust the wheels until the active-door panel is parallel to the jamb. At the same time I make sure the door still slides smoothly. Before trying to turn the adjustment screws, I pick up the door a little to relieve some of the weight on the wheel,

Most panels go in top first. The door panels are installed after the frame is secure in its opening. Most door panels slip into a channel in the head jamb first; then the bottom edge is set down over the track on the sill.

which is especially helpful with some of the larger, heavier door panels. While adjusting the wheels, I also keep an eye on the alignment at the center, making sure the stile of the sliding-door panel is even and parallel with the fixed panel and that the muntin bars, if there are any, stay lined up.

When I'm satisfied with the way the panels are set, I drive home the screws in the side jambs, placing additional shims behind the strike location. To get the head jamb perfectly straight, I stretch my string and shim the head down to the string. I then secure it with screws or nails.

A Moisture Barrier Seals the Outside of the Slider

In many parts of the country, door installers run strips of felt paper under the nailing flanges on the side jambs to weatherproof sliding doors. Here in California, where stucco is the most-common siding material, the seal around the slider definitely has to be waterproof. So we use a system that is recommended by many manufacturers as well as by the National Fenestration Rating Council.

After the nailing flanges have been bedded in caulking against the sheathing, we run another bead of caulking on top of the nailing flanges, and a layer of Moistop is pressed into the caulk. The side jambs are flashed first, with the Moistop extending at least 8 in. above the head jamb. I then apply a length of paper flashing across the head jamb, again pressing it into a bead of caulk on top of the nailing flange. The flashing on the head jamb should be long enough to overlap the flashing on the side jambs.

In areas exposed to extreme weather, such as places near the ocean, we follow the same steps to waterproof doors and windows except that we use self-healing, adhesive-backed waterproof membrane instead of paper flashing.

Tuning the sliding panel. A screw is turned to raise or lower the wheel, which brings the edge of the sliding panel parallel to the jamb.

Screens Should Slide as Smoothly as Main Panels

Putting in the screens is probably the most frustrating part of sliding-door installation. If manufacturers were graded by the quality of their screen doors, a lot of them would fail miserably. The worst screen doors have little plastic wheels without bearings. Often, these wheels cannot be adjusted and instead rely on a spring to counterbalance the door. When you try to slide the door, it acts like a rocking horse, dragging first on the sill, then on the head.

I always install screens top first into the screen channel in the head. While lifting the door against the head jamb, I raise the screen over the sill and set the bottom wheels down on top of the track. If the bottom wheels don't clear the track, I engage the top of the screen, then gently set the screen down next to the lower track. I then lift the screen from the bottom and push each wheel up with a stiff putty knife guiding it onto the track (see the left photo below). The difficulty of screen installation seems to increase in direct proportion to the amount of wind. So pick a calm day or get an extra pair of hands to help on windy days.

Most screens have wheels that adjust up or down by turning a screw in the leading or trailing edge of the door (see the right photo below). If the sill has been installed level and straight, it usually takes just a slight adjustment to get the leading edge of the screen parallel to the jamb. If there are wheels on top as well, I reduce their tension until the door slides smoothly.

The best sliding screens I've seen are on Marvin® doors (Marvin Window and Door, PO Box 100, Warroad, MN 56763; 800-346-5128). Marvin screens are suspended from an upper track, and the spring-loaded wheels on the bottom just keep the bottom of the screen in line. I place the Marvin screen door first on the lower track, and I make sure the door slides easily on its lower wheels. Then I climb a ladder and slip the upper wheels into the upper track. All the adjustments are made on the upper wheels by turning a single cam-action screw.

By their nature, slider screens tend to be flimsy, so it's not uncommon to see them come from the factory slightly out of square. An out-of-square screen usually doesn't glide smoothly and is difficult to line up with the jamb. But in most cases these parallelogram screens can be squared by applying diagonal pressure on the screen frame. If the wheels

A putty knife puts screen wheels on track. If the screen panel rides on spring-loaded wheels, a stiff putty knife can be used to depress the wheel and guide it onto its track.

Screws adjust the screen wheels. Screws in the edges of the screen adjust the height of the spring-loaded wheel as well as the tension on the springs to keep the screen from binding.

on the screen are not adjustable, racking the screen is often the only way to make it line up with the jamb.

Screen doors on wood sliders have bumpers to keep them from damaging the jambs. But if the screen has just a single bottom bumper, the wheels have a tendency to bounce off the track when the bumper hits the jamb. I make sure that there are bumpers at both the bottom and the top of the screen door so that the back of the screen hits the jamb evenly. And with all sliding screens, a little silicone spray on the track helps keep those wheels gliding smoothly.

Installing the Locks and Hardware

Most sliders come from the factory prebored for their latch hardware, although a few come with the strikes installed. Installing strikes is one area in which reading the instructions and mocking up the hardware really helps. The strikes have to be located after the doors and latches are installed.

Some latches, especially those on screen doors, have a jaw or hook that engages the strike when the latch is closed. To locate the strike for this type of latch, I extend the latch mechanism and slide the door up to the jamb. I make a mark on the jamb in line with the bottom of the latch jaw (see the near right photo at right), which is the location of the proper position for the opposing jaw on the strike.

Other slider locks are internal, and the strike seats inside the latch. To line these up, I engage the strike in the latch and slide the door up against the jamb. A sharp point on the backside of the strike marks the jamb to locate the proper position for the strike (see the far right photo at right). Some strikes have to be mortised into the jamb for the door to close properly. But I always install the strike on the surface of the jamb first to figure out how deep I need to make the mortise. I try not to mortise the strike too deep or to close the gap between the slider

Locating the screen strike. For most screens and many sliding doors, the strike is located by sliding the panel over and making a mark in line with the latch jaw.

A point locates this strike. If the jaw is located on the strike instead of the latch, the strike is engaged in the latch, and the panel is slid to the jamb. A point on the back of the strike marks its location.

and the jamb entirely. Adequate room has to be allowed for the weatherstripping, and too much pressure can jam the lock.

Speaking of pressure, you're probably wondering what happened at that lady's house. I discovered that the sill was sitting on a foundation anchor bolt that I was able to trim off with my reciprocating-saw blade slid under the sill. I needed to draw the oak sill down, so I drilled through it in three spots, first counterboring for wood plugs. I used concrete screws to pull down the sill, an easy way to go if you can't remove the sill to install anchors. Three screws sucked that sill right down, and adjusting the doors was easy after that. The lady thought I was a genius.

Gary M. Katz is a carpenter in Reseda, California, and the author of The Doorhanger's Handbook *(The Taunton Press, 1998).*

Cutting Out
Basic Stairs

■ BY ERIC PFAFF

started out as a laborer on a framing crew ten years ago. After a few months I began feeling cocky about my framing skills and one day decided that I was going to impress the boss. I asked him if I could cut a simple set of basement stairs while he was on lunch break. He told me to go for it, though he wasn't going to pay me extra for working through lunch. Anxious to prove myself, I agreed to his terms.

I worked frantically, first figuring the rise and run, then cutting the stringers as fast as I could. I was just nailing the last of the treads when the boss came back from lunch. He seemed surprised that I had gotten the steps done, but the real surprise came when he smacked his forehead on the stairwell header. In my haste I had miscalculated the headroom over the stairs and never got a raise while working with that crew.

Measure the Height of the Stairs Where They Land on the Floor

I've cut over 100 sets of steps since I gave my boss that concussion, and I now realize that the first and most important step in stair building is accurately calculating the rise, or the height of each step, and the run, or the width of each step. I begin by finding the overall rise, or the distance between the two floors that the stairs will connect.

In a perfect world all floors would be flat and level, and measuring the rise would mean simply running a tape from the floor above to the floor below. However, I've seen floors— especially in basements—that slope a couple of inches from one wall to the other.

The first and most important step in stair building is accurately calculating the rise, or the height of each step, and the run, or the width of each step.

Basic Stair Terminology

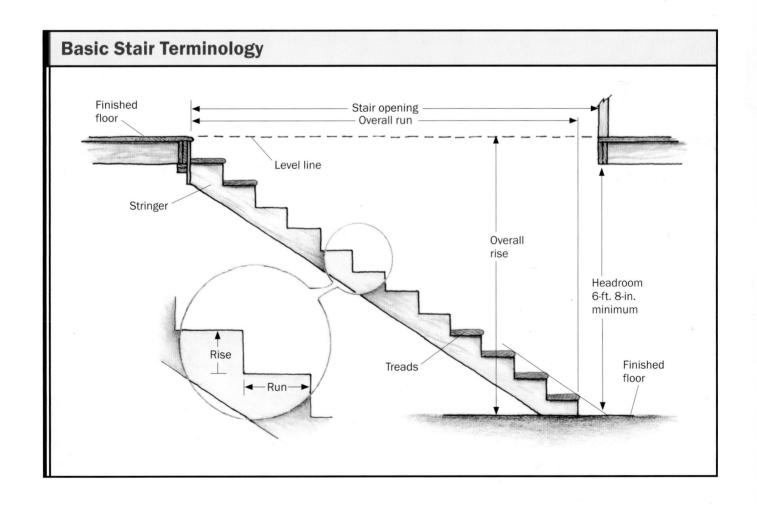

Finished floor

Stair opening
Overall run

Level line

Stringer

Rise

Run

Overall rise

Headroom 6-ft. 8-in. minimum

Treads

Finished floor

The way around this problem is measuring the overall rise as close as possible to where the stairs will land on the floor. I make this measurement by taping my 4-ft. level to a straight 2x4 that is at least as long as the framed opening for the stair (see the photo below). Keeping the 2x4 level, I measure up to it near to where I figure the stairs will land. In the stair featured here, the overall rise is 99½ in.

Measuring the overall rise. A 2x4 extends a 4-ft. level over to where the staircase will land. Measuring the rise at this point minimizes the chance of error from a basement floor that might not be level.

Figuring Rise and Run

Many codebooks and safety officials insist that all stairs have a 7-in. rise and an 11-in. run. However, the formulas below are also routinely used to determine rise and run.

1. The rise times the run should equal approximately 75 in.
2. Two times the rise plus one run should equal 25 in.
3. Rise plus run should be 17 in. to 18 in.

Check Local Codes before Cutting Stairs

Before I explain my calculations, let me say a brief word about stairs and the codebook. Code requirements for stairs seem to change with every new codebook and can vary greatly from state to state, sometimes even from town to town. For instance, some codes require a 7-in. maximum rise; others allow an 8-in. rise. So check with your local building inspector to make sure that any stairs you build meet the local code. For the project featured here, I was replacing an existing set of basement stairs. My floor heights were fixed, as was the rough opening in the floor, so I had to work within the constraints of the existing framing. Consequently, these stairs are steeper than most codes allow. But the building code in use at the time they were built allowed exceptions when replacing existing stairways.

Generally, I think a 7½-in. rise with a 10-in. run produces a safe, comfortable stair, so as a starting point, I divided my overall rise by 7.5, which gave me 13.266 rises. You can't have a partial step, so I divided the total rise by the nearest whole number, in this case 13. The result is 7.653, or very close to 7¹¹⁄₁₆ in. for each individual rise.

Now that I know how many steps I'll have, I can figure out the depth, or run, of each tread. The header that will support the staircase at the top of the stairs will act as the first riser, so the stringers actually end up with 12 risers and 12 treads. (In every stair, you'll have one less tread than the number of risers.) For maximum headroom I'd like the bottom step to land directly below the other end of the framed opening, which is 108 in. long. So I divide 108 in. by 12, the number of treads, and end up with a tread depth of 9 in. As mentioned previously, code requirements for stairs may vary, as well as the formulas for calculating safe and comfortable stairs (see the sidebar at left). My local building officials tell me that the rise plus the run of a stairway should be 17 in. to 18 in. At 16¹¹⁄₁₆ in., this staircase is a bit steeper than I'd like. Again, I'm restricted by the existing framing, but I'd rather have a slightly steep stair than compromise the headroom clearance. For this project, the inspector agreed with me.

Stair Gauges Help with the Stringer Layout

A stringer, or carriage, is the diagonal framing member that holds the treads (and the risers, if they're used). I like to use straight, kiln-dried Douglas fir 2x12s for stringers whenever possible. I start by setting a 2x12 on sawhorses with the crown, if any, facing toward me. A framing square is the best tool for laying out the sawtooth pattern of treads and risers on the stringer. But I also use stair gauges, which screw onto the framing square and increase the speed and accuracy of the layout (see the top photo at right). Available at most hardware stores, these little beauties are small hexagonal blocks of aluminum or brass with a slot cut in them so that they slip over a framing square. Thumb screws or knurled nuts hold them in place on the square. I put one gauge at the rise number on the short side of the square (the tongue) and the other at the run on the long side of the square (the blade). The gauges register against the edge of the 2x12 and keep the square orientation on the stringer consistent for every step.

The Thickness of a Tread Is Subtracted from the Bottom of the Stringer

As I move down the stringer and lay out the steps, I line up the edge of my rise with the run line from the step above. I repeat

Stair gauges streamline layout. Small clips called stair gauges are attached to the framing square at rise and run measurements to keep the orientation of the framing square the same for each step layout.

Subtract the thickness of a tread. For the bottom rise to be the correct height, the thickness of one tread must be subtracted from the bottom of the stringer. Moving the square to the other side of the stringer makes it easier to complete the lines.

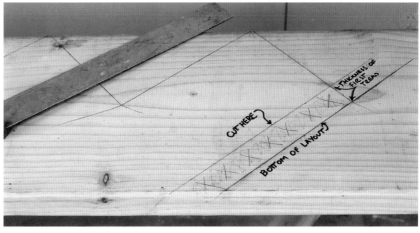

Label lines to avoid confusion when cutting the board.

the process all the way down the stringer until I have the right number of rises, in this case 12.

After I've drawn the last rise on the bottom of the stringer, I square it back using my framing square (see the middle photo on p. 229). If the stringers were cut and installed and the treads put on at this point, the bottom step would be higher than the rest of the steps by the thickness of the tread. I solve this problem by subtracting that thickness from the bottom of the stringer, which lowers the stringer assembly. That way, when the stair treads are installed, all of the rises will be the same.

As a word of caution, the overall rise of the stair should be calculated from the finished floor above to the finished floor below. If you didn't compensate in your original measurement, adding finished flooring after stairs are built will result in a top or a bottom step different from the rest, which creates a dangerous stair.

Because my treads will be made out of 2x10s, I measure up 1½ in. from the bottom of the stringer. I always label my lines to avoid confusion when cutting (see the bottom photo on p. 229).

First cuts are made with a circular saw. Overcutting the lines will weaken the stringer, so the initial cuts with a circular saw should never extend beyond the layout lines.

Finishing the cut. A jigsaw is used to complete the cuts in each corner, cutting the wood that the circular saw can't get. A handsaw will also work instead of a jigsaw.

Overcutting Can Weaken the Stringer

After laying out the stringer, it's time for me to cut the rises and the runs. I make my initial cuts with a circular saw, making careful, steady cuts for all of the rise and run lines along the length of the stringer. I take care not to overcut the lines, which would substantially weaken the stringer. Instead, I finish my rise and run cuts with a jigsaw (see the bottom photo at right).

After the stringer is cut, I set it in place temporarily to make sure that it fits and that it's level. I figure out exactly where the stringer will rest against the header of the stair opening by measuring down from the floor the distance of the rise plus the thickness of the tread ($7^{11}/_{16}$ + $1½$ = $9^{3}/_{16}$ in.).

I make a level line across the header at this measurement and tack my stringer in place so that the top of the stringer is even with this line. For this stairway I had to extend the header down with a 2x4 and plywood to give me more to attach the stringer to.

Next I test several of the steps with a torpedo level. Once I'm satisfied with the stringer, I take it down and use it as a template to lay out and cut the second stringer (see the top photo on the facing page). I make sure that the sawtooth points on the template stringer point in the same direction as the crown of the second 2x12.

The second stringer is cut the same as the first, and then both stringers are tacked in place and checked for level. This time I check the runs on the individual stringers for level and also use my 4-ft. level to make

The first stringer becomes a template. Once the first stringer has been cut and tested in place, it can be traced for the layout of the second stringer. The crown of the stringer stock must be facing the same direction as the points on the template.

Not only are railings an essential part of stair safety, they are also the law.

sure they are level with each other (see the photo at right). If a staircase with 1½-in.-thick treads is more than 36 in. wide, it's necessary to use three stringers. Again, it's best to check with your local building official if you have any doubts or questions.

Strongbacks Take the Flex out of the Stringers

Before I install the stringers permanently, I add strongbacks to each one (see the top photo on p. 232). Strongbacks are 2x4s nailed at right angles to the stringer and run their entire length. I nail the strongbacks through the stringers with 16d nails every 8 in. Putting them on the outside of the stringers lets the stringers be closer together, which reduces the span of the treads and makes the stairs feel more stiff. Strongbacks stiffen the stringer by limiting lateral flex.

After the strongbacks are nailed on, I install the stringers for good. I nail them securely at the top through the plywood header extension from behind and into a 2x4 spacer block between the stringers. At the bottom a length of pressure-treated 2x

Both stringers are tested together. Before the stringers can be installed, both are set up and leveled individually. A 4-ft. level then tests them in relation to one another.

Strongbacks stiffen the stringers. Strongbacks, or 2x4 stiffeners, are nailed to the sides of the stringers to keep them from flexing.

Treads are attached with screws. Screwing the treads to the stringers is the best way to keep the staircase from squeaking. Predrilling the holes prevents the screws from snapping while they're being driven.

nailed to the concrete with powder-actuated fasteners serves as a nailer as well as a spacer between the bottom ends of the stringers. Some builders like to notch the bottom of the stringers around the nailer. But with the bottom step already cut 1½ in. narrower for the thickness of the tread, notching out for the nailer leaves a thin and precarious section of diagonal grain that is more likely to split when the treads are attached or to break off if I trip over the stringer before the treads are installed.

The final step is putting on the treads. I cut the treads for this stairway out of 2x10s and attach them to the stringers with 3-in. screws. Screws are the best way to keep the treads from squeaking, and if there is ever a problem, they can be removed quickly without damaging either the tread or the stringer. I predrill my holes, especially if I'm using bugle-headed construction screws, which can snap as they're going in.

One important item that I'm obviously leaving out is the railing system. Not only are railings an essential part of stair safety, they are also the law. This stairway will be enclosed with a wall on the open side, and handrails will be hung on each wall to make the stairway complete. Stair railings are a topic for another chapter.

Eric Pfaff is founder and principal of Architectural Automations, which automates multi-sliding doors. He is the author of The Quicky Stair Book.

Disappearing Attic Stairways

■ BY WILLIAM T. COX

When I was young and my mother wanted something out of the attic, she would push me up a stepladder and through a little access hole in the ceiling; it was a scary adventure for an 8-year-old, climbing up into a dark, cavelike hole where I thought unknown creatures waited to devour me. What we needed was a disappearing stairway.

Disappearing stairways are available in several styles. All of these stairways have a ceiling-mounted trap door on which the stairway either folds or slides. Nearly all are made of southern yellow pine, although there are a few aluminum disappearing stairways. There are a few commercial models made of aluminum or steel, but this chapter will concentrate on residential models. Disappearing stairways are not considered to be ladders or staircases, and they do not conform to the codes or the standards of either. Disappearing stairways have their own standards to which they must conform.

Similar to ladders, disappearing stairways have plenty of labels and warnings to read. On all disappearing stairways there are warnings about weight limits because, inevitably, homeowners fall down stairs while trying to carry too much weight into the attic. Also, labels tell the user to tighten the nuts and bolts of the stairway.

Sliding stairway. Disappearing stairways are concealed by a spring-loaded ceiling door. Here, the author walks up a sliding stairway made by Bessler with an angle of incline close to that of a permanent stairway.

In one stairway manufacturer's literature, the word "safer" was used to describe the fluorescent orange paint used on the stairway's treads. But "safer" was replaced by "high visibility" because one homeowner wore off the paint, slipped, and fell. She sued both the manufacturer and the builder because, she claimed, the treads became unsafe to use.

Stairway companies are constantly testing, upgrading, and improving their products to give the consumer the best, safest, and longest-lasting disappearing stairway possible. And with good reason—over a million units were produced in the United States last year.

Folding Stairways

The most popular style of disappearing stairways, folding stairways consist of three ladderlike sections that are hinged together, accordion style. The three sections are at-

tached to a hinged, ceiling-mounted door similar to a trap door. The door and the attached ladderlike sections are held closed to the ceiling by springs on both sides. When you want to access a folding stairway, you pull a cord that is attached to the door and lower the door from the ceiling. The door swings down on a piano hinge. You then grab the two bottom sections of the stairway and pull them toward you, unfolding them (see the photo below). When the two bottom sections are completely unfolded, all three sections butt together at their ends, giving strength and stability to the stairway.

Folding attic stairs are measured by the rough opening they occupy and by the floor-to-ceiling height they will service. The smallest folding stairways are 22 in. wide, and they are made to fit between joists 2 ft. o.c. These narrow stairways are available in models that will service a ceiling height as short as 7 ft., and there are others that can go as high as 12 ft. Keep in mind that a stairway's rough-opening width is appreciably more than the actual width of the ladderlike sections. Because of the attendant jambs, springs, and mounting hardware necessary to operate the stairway, the actual width of the ladderlike section is a lot less than the rough opening. A stairway with a rough opening of 22 in. is going to have a tread about 13 in. wide.

Folding stairways are rated according to weight capacities; the lightest-rated ones will handle 225 lb., and most of the others have a recommended weight capacity of 300 lb. It is interesting to note that American Stairways, Inc., says in its product literature that you are not supposed to carry anything up or down its stairways. Only an unladen person is supposed to climb up or down. This all sounds somewhat ridiculous to me. The reason why people install disappearing stairways is that they can carry stuff up or down from the attic—Christmas ornaments, baby clothes. However, I tell customers not to carry stuff up the folding stairway. You should have someone hand it up to you. You can-

Folding stairway. Ladderlike sections are hinged like an accordion to the ceiling-mounted door. On this model, made by American Stairways, the treads are painted with bright-colored, rubberized paint.

not climb a folding stairway with something in your hands. It's way too steep. I suppose the disclaimer keeps American Stairways out of court if somebody falls down one of the stairways. Also, all folding stairways are for residential use only; a restaurant owner once asked me to install a folding stairway so that he could access a storage area above the kitchen, and I had to refuse.

The smallest folding stairway costs around $75, and the largest, the A-series aluminum folding stairway made by Werner®, costs around $211. It fits a rough opening of 2 ft. 1½ in. by 4 ft. 6 in. and accommodates a ceiling height of up to 12 ft.

Installing Folding Stairways

Aside from the finish trim, folding stairways come out of the box as a complete, assembled unit. (Other types of stairways require some assembly.) Because most of the installations I do are retrofits into existing buildings, the first thing I must do is cut a hole in the drywall. If possible, I try to mount the stairway alongside an existing joist; this saves some framing work if the stairway is bigger than the space between two joists. Cutting the drywall is not a close-tolerance operation because (within reason) the finish trim will cover any ragged edges. If I have to head off a ceiling joist, I use standard carpentry practices.

Here's a time-saver I came upon after installing quite a few stairways. I've found that it's much easier to cut and fit (but don't nail) the finish trim while the stairway is sitting on the floor in front of me rather than on the ceiling. Leave ⅛ in. between the edge of the door and the jamb. Make sure you mark the location of all four pieces. Once the stairway is installed, you just nail the pieces in place.

Before installing the stairway, I screw two temporary ledgers to the ceiling that project ¾ in. into the rough opening. The ledgers provide a shelf for the stairway's

wood frame once I've lifted the unit into the rough opening (see the top photo below). When attaching the ledgers, I make sure they are parallel and that they only stick into the opening ¾ in. Any farther than that, and they might not allow the door to swing open on its piano hinge. Using screws instead of nails to attach the ledger makes it possible to adjust them in case I somehow miscalculate; it also makes them easier to remove when the time comes.

Support during installation. Ledgers screwed to the ceiling provide temporary support for the stairway while the author shims and screws the frame to the rough opening.

An extra screw for insurance. A third mounting screw in a folding stairway's piano hinge strengthens the installation. The author drills through the hinge and the jamb and into the rough framing. The large spring at the top of the photo is one of a pair that holds the stairway and its trap door closed to the ceiling.

TIP

It's much easier to cut and fit (but don't nail) the finish trim while the stairway is on the floor rather than on the ceiling. Leave ⅛ in. between the edge of the door and the jamb. Make sure you mark the location of all four pieces. Once the stairway is installed, just nail the pieces in place.

Measuring for trimming. With the bottom section folded under the middle section, the author puts his weight against the stairway to ensure it is fully extended. He measures along both the top and bottom edges of the stairway, transcribes his measurements on the bottom section, connects the dots and makes his cut. The trap door's pull cord can be seen hanging in the top of the photo.

Although some manufacturers warn against it, I usually remove the bottom two ladderlike sections of the stairway before carrying it up the stepladder. Most often I work alone, and some of the stairways are pretty heavy to lift by myself. A 30-in. by 54-in. stairway made by American Stairways, Inc., weighs 92 lb.

With the ledgers in place, I lift the stairway into the rough opening and set it on the ledgers. Next I carefully open the door fully and center the jamb in the rough opening. Now it's just a matter of shimming the sides of the frame and fastening them to the framing. (Once the unit is installed, I reattach the sections and tighten the nuts and bolts on the hinges with a screwdriver and a socket wrench.)

Most instructions call for nailing the frame in place, but I like to use screws because they are more adjustable than nails,

and they are also easier to remove if needed. I start at the hinge end of the stairway jamb. Most folding stairways have a hole drilled at both ends of the piano hinge to screw the hinge into the framing. I always drill another hole through the hinge and sink a third screw (see the bottom photo on p. 235). I use #10, 3-in. pan-head screws. Adding a third screw can't hurt, and it only takes an extra minute or two.

Instructions call for screwing or nailing into the framing on both sides of the stairway through two of the holes drilled in the arm plate, which is the metal plate to which the door arms are attached. I shim behind the arm plates because it is critical that the arms stay parallel to the ladder and that the pivot plates remain stationary. If they don't, the rivets that hold the arms will wear out from twisting and torquing as the stairway is used.

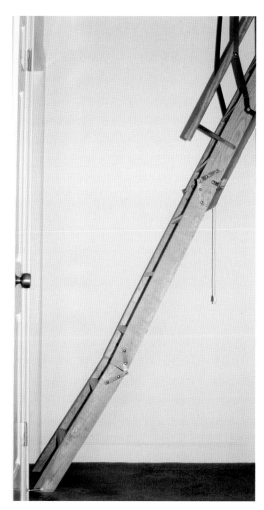

Accurate cuts are important. A folding stairway that is cut too long puts undue stress on the hinges because the ladderlike sections don't butt at their ends.

After I've screwed through the piano hinge and the arm plates, I shut the door and make sure there is an even reveal between the door and the jamb all the way around the door. When this is done I shim and screw off the rest of the wood frame, using #8, 3-in. wood screws.

Cutting Stairs to Length

Because ceiling heights vary, folding attic stairways come in different lengths, and with the exception of aluminum models, you must cut the bottom ladderlike section to length when installing the stairway. It is not difficult to figure out the cut length, but it is critical to the longevity of the stairway that the length be exact. A stairway that is cut too long will not extend to a straight line, and the ends of the ladderlike sections will not butt together (see the photo at left), putting undue stress on the hinges. A stairway that is cut too short will stress the arm plates, the counterbalancing springs, and the section hinges.

To cut the bottom ladder section to length, I make sure the arms are fully extended and fold the bottom section underneath the middle section. I rest my leg against the stairway to ensure that it is fully extended, and I take my tape measure and hold it along the top, or front, edge of the middle section (see the photo on the facing page). By extending the end of the tape to the floor (while holding the upper part of the tape against the middle section), I get an exact measurement from the floor to the joint between the two lower sections. I repeat the procedure on the back edge of the stairway to get the length of the back of the cut. Then I remove the lower section, transcribe the measurements and draw a line between the two points on each leg.

After making my cuts and reattaching the bottom section of the stairway, all that's left to do is unscrew the temporary ledgers from the ceiling and run the precut trim around the frame. I've installed quite a few folding stairways, and I can usually manage to do the whole job in about two hours.

Sliding Stairways

Several companies make folding stairways, but Bessler Stairway Company also makes a sliding disappearing stairway. Unlike a folding stairway, where the sections are hinged and fold atop one another, the sliding stairway is one long section that slides on guide bars aided by spring-loaded cables mounted in enclosed drums. When the stairway is closed, the single-section stairway extends beyond the rough opening into the floor

Sliding stairways, unlike folding stairways, are designed so that the user can walk up into the attic while carrying a load.

space above. This is an important consideration because some small attic spaces do not have enough room for the stairway's sliding section.

To access a sliding stairway, you simply pull the door down from the ceiling, similar to the way you'd pull down a folding stairway. Then you grab the single ladderlike section and slide the section toward you, lowering it to the floor. To close the stairway, you slide the single section back up into the opening. A unique cam-operated mechanism locks the ladderlike section in place while you push the door back to the ceiling. A series of spring-loaded, counterbalancing cables makes the door and the ladderlike section feel almost weightless.

The real benefit of sliding stairways is their angle of incline. Folding stairs typically have about a 64-degree angle of incline. That's pretty steep—more like a ladder than a staircase. Bessler's best sliding stairways have a 53-degree angle of incline. Sliding stairways, unlike folding stairways, are designed so that the user can walk up into the attic while carrying a load (see the photo on p. 233).

Sliding stairways are made of knot-free southern yellow pine, and there are four different models from which to choose. The smallest—the model 26—has a rough opening of 2 ft. by 4 ft. and has a suggested load capacity of 400 lb. This model has a stairway width of 17$\frac{1}{16}$ in. The model 100 requires a rough opening of 2 ft. 6 in. by 5 ft. 6 in. and has a suggested load capacity of 800 lb. The width of the stairway is 18$\frac{7}{8}$ in. Sliding stairways are measured from floor to floor, rather than from floor to ceiling like folding stairways, and the largest model 100 will service a floor-to-floor height of 12 ft. 10 in. Sliding stairways also have a full-length handrail.

The smallest sliding stairway, the model 26 with a maximum ceiling height of 7 ft. 10 in., costs around $225. The largest model, the model 100 with a maximum ceiling height of 12 ft. 10 in., costs around $700.

Installing Sliding Stairways

Sliding stairways do not come from the factory as assembled units; installation of these stairways is more for a journeyman carpenter because the finished four-piece jamb is not furnished and must be built on site. Stringers and treads need assembly, and the door and all hardware have to be installed on site.

I frame the rough opening 2 in. larger than the door opening. This allows me to use ¾-in. stock for the jamb and still have ¼ in. of shim space on each side to account for possible framing discrepancies. I rip the jamb stock to a width equal to the joist plus finished ceiling and attic flooring material.

It's possible to attach the finish trim to the jamb while it's still on the floor and then mount the whole unit into the rough opening using braces (called stiff legs or dead men) to hold the jamb to the ceiling while its being shimmed and nailed. But because I work alone, I screw ledgers to the ceiling the same way I do for folding stairs and then apply the trim later.

I nail the hinge side of the jamb to the rough framing and then hang the door with #10, 1-in. pan-head screws. Next I close the door to fine-tune the opening. After eyeballing the crack along the door edge, I move the jamb in and out to produce an even reveal down each side and then shim and nail the jamb.

Next I lay the stringers on sawhorses and thread the ladder rods with washers through the center holes of both stringers so that the stringers will stand on edge. Ladder rods are threaded rods that go under the wood treads, giving strength and support to the sections. I install all but the top three treads into the gains (or dadoes) in the stringers, screw the treads to the stringers, then tighten the nuts on the ladder rods. I always peen the ends of the ladder rods to keep the nuts from falling off. It's important to leave out the top three treads so that I can slide the ladderlike section onto the guide-frame bars at the top of the finished jamb.

When I install the guide frames and the two mounting brackets for the drums that contain the springs, I always predrill all of the holes with a %4-in. bit. After 30 years, you would be amazed to see how the wood pulls away from where the screws were put in without predrilling. This causes a minute split to start, and when I repair sliding stairs that are 30 to 50 years old, the cracks have grown enough that I can stick a finger into them.

Installation of the mounting hardware is pretty straightforward. After putting the stringers onto the guide bars, I attach the cables. Caution: I wear gloves and am careful adjusting the cables' tension around the drums. If the cable slips, I could wind up like in *The Old Man and the Sea*, with deep cuts in my hands and no fish dinner.

William T. Cox is a carpenter in Bethel Springs, Tennessee, who specializes in installing and repairing disappearing stairways. He is building an earth shelter.

Sources

American Stairways, Inc.
3807 Lamar Ave.
Memphis, TN 38118
901-795-9200

American makes three models of folding disappearing stairways. The smallest has 1x4 treads and stringers and a rough opening of 22 in. by 4 ft. The largest has 1x6 treads, 1x5 stringers, and a rough opening of 2 ft. 6 in. by 5 ft. Scissor hinges join the ladderlike sections. Optional accessories include an R-6 insulated door panel, bright orange rubberized painted treads, and a fire-resistant door panel.

Bessler Stairway Co.
3807 Lamar Ave.
Memphis, TN 38118
901-795-9200

Bessler is a division of American Stairways, Inc. Bessler makes a folding stairway as well as a sliding stairway that has a one-piece stringer and slides on guide bars counterbalanced by spring-loaded cables. Bessler's folding stairway has high-quality section hinges that butt when the stairway is opened. Standard features include 1x6 treads and 1x5 stringers, as well as

an R-6 insulated door and bright orange rubberized painted treads.

Memphis Folding Stairs
PO Box 820305
Memphis, TN 38182
800-231-2349

Memphis Folding Stairs makes folding stairs that are very similar to the ones offered by American and Bessler. In fact, a person who worked for Memphis Folding Stairs now owns American Stairways. They also sell an aluminum folding stairway, as well as a heavy-duty wood model with 2x4 rails and 2x6 treads. Memphis sells a thermal airlock for its stairs that covers the stairway opening. It operates like a roll-top desk and has an R-value of 5.

Precision Stair Corp.
5727 Superior Dr.
Morristown, TN 37814
800-225-7814

Precision makes metal folding stairways and a fixed aluminum ship ladder with a 63-degree angle of incline. The company also makes an electrically operated commercial-grade sliding stair that has a switch at both the top and bottom of the stairway.

R. D. Werner Co., Inc.
93 Werner Rd.
Greenville, PA 16125
724-588-8600

R. D. Werner is a large ladder manufacturer that also makes the Attic Master, which is its line of folding stairs. Of particular note is its aluminum stairway with adjustable feet and a load capacity of 300 lb. Options include a wood push/pull rod that takes the place of a pull cord, self-adhesive antislip tread tape, and a stairway door R-5.71 insulating kit.

Therma-Dome, Inc.
36 Commerce Cir.
Durham, CT 06422
860-349-3388

Therma-Dome offers two insulating kits for attic stairs (R-10 and R-13.6) that consist of foil-covered urethane foam boards and touch-fastener tie-downs. These covers seal to the attic floor with a foam gasket. With their high R-values, payback will be quicker in colder climates. The covers cost between $65 and $80. Therma-Dome will fabricate covers for most stairways.

Installing Handrails from Stock Parts

■ BY LON SCHLEINING

Of all the parts that make up a home's interior, the most visually dramatic is often the staircase. In particular, a stair's handrail is a combination of precise joinery and crisply milled components that looks as delicate as it is strong. It should come as no surprise, then, that building a handrail can be a complicated job, even for an experienced carpenter.

After installing hundreds of handrails, I've come up with a pretty simple system to handle everything from manufactured stock parts to custom stair railings. On a recent remodel, I installed new turned newel posts, handrails, and balusters on a staircase that would be carpeted later. These few examples won't be the final word on how to install handrails, but they should get novices off to a solid start and perhaps offer more experienced carpenters a new trick or two.

A Full-Size Drawing Is the First Step to Easier Railing Installation

Nearly all the instructions I've seen describe fitting a handrail by first laying it on the stairs instead of setting it on the previously installed newel posts. The installer attempts to clamp fittings in place or uses charts and formulas to determine the height of newels; easings (such as volutes and goosenecks) are trimmed with the aid of a pitch block.

Instead, I find that making full-scale drawings is the easiest and fastest way to lay out handrails (see the photo on the facing page). It's much easier working through difficult problems on paper than making mistakes with expensive stock. I use 60-in.-wide brown wrapping paper from Papermart (800-745-8800) for these drawings.

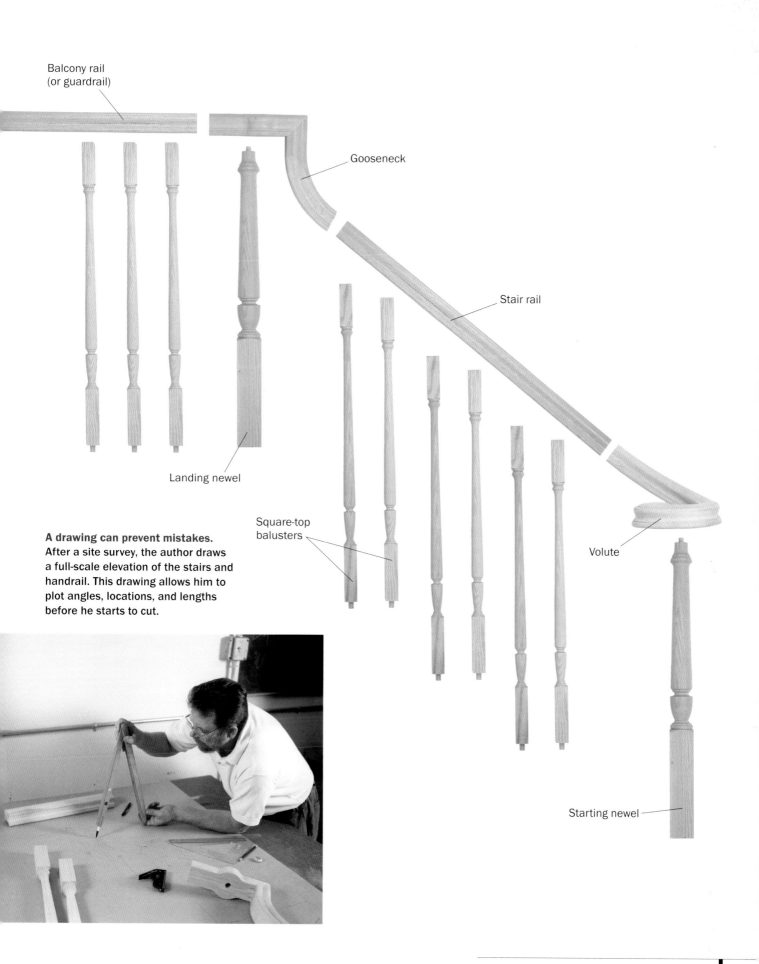

Balcony rail
(or guardrail)

Gooseneck

Stair rail

Landing newel

Square-top
balusters

Volute

A drawing can prevent mistakes.
After a site survey, the author draws
a full-scale elevation of the stairs and
handrail. This drawing allows him to
plot angles, locations, and lengths
before he starts to cut.

Starting newel

After taking the necessary measurements at the job site, I draw the top, the bottom, and an intermediate step in elevation view to locate top and bottom newels. Obviously, there are usually more steps than that in a single run, but I can figure the pitch angle of the stair and handrail by drawing a line touching the three tread nosings. Next I draw the handrail at its correct angle and height determined by code, in my case between 34 in. and 38 in. above the nose of the tread. If there is a guardrail, I draw it at least 36 in. above the floor. I draw certain areas, such as the starting step, in plan view as well as elevation to locate the volute position.

In a few minutes, I'm able to make a list of parts that I need and do nearly all the planning for the whole job. When I get to the site, I unroll the drawing and refer to it for layout work. It's also a great device for discussing designs with the clients.

At this stage, I make sure that the staircase will pass muster with the building inspector. For instance, the building code in California specifies a maximum opening of 4 in. between balusters, so I may need to draw three balusters per tread instead of the more common two per tread I might have planned.

Quality Parts Are a Must

After drawing the plan, my next step is to buy parts. Stock parts from larger manufacturers will almost always be usable, but you or your client might be more observant than the factory inspector. Cosmetic flaws such as checking, grain tearout, color variations, obvious chatter marks, and uneven machining will really stand out on a finished staircase. It's always a good idea to look before you buy; if you can't, ask about the manufacturer's (or retailer's) exchange policy. I buy most of my parts from Leeper's Wood Turning (800-775-1173). They're local for me (they're also a mail-order retailer), and they offer good quality. I've also had good results

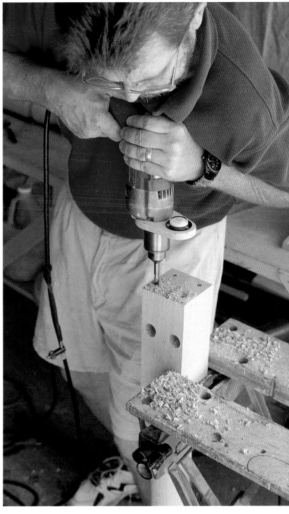

Bubble level keeps the drill straight. After drilling access holes in the newel, the author clamps the post plumb and drills four bolt holes that must be true. The bubble level's plywood base fits tightly onto the drill's motor housing.

with parts from L. J. Smith® (740-269-2221) and Coffman® (540-783-7251), two manufacturers that distribute nationwide.

Handrails, probably the most scrutinized part of a staircase, are available in solid or finger-jointed stock. Some people don't like finger-jointed stock because the joints become highlighted when the railing is stained, so I always check with the client first. When ordering, I specify lengths that will span between the newels on the stairs and balconies, assuming that a certain amount will be trimmed off for fittings. I make sure to draw the balusters on my full-size layout sheet. Although adjacent balusters on a tread are not all the same length, each should show a consistent proportion of turning. I also buy extra hanger bolts, washers, nuts, screws, and plugs, just in case.

Strong Railings Need Sturdy Newels

Starting and balcony newels are typically bolted to the framing. If the newels are not securely fastened, the staircase might look pretty, but the railing won't be strong. To make certain that it's solid, I always build the starting step from medium-density fiberboard that's bolted to the floor. After cutting the newel to length and marking its location on the step, I lay out the four bolt holes, both on the newel end and on the starting step. (Not everyone uses four bolts in this situation, but I think four bolts give you

more adjustability and keep the post from twisting.) With the newel clamped securely, I use a Forstner bit to drill the 1-in.-dia. access holes, then drill four ⅜-in. holes in the post end to a depth of about 4 in.; here, it's important to drill straight (see the photo on the facing page).

After drilling pilot holes, I run four ⁵⁄₁₆-in. by 6-in. hanger bolts (those with wood threads on one end and machine threads on the other) into the step, and then slip the newel onto the bolts (see the top left photo below). To start the nuts deep inside the access holes, I use a cool gizmo (see the bottom left photos below) made by Universal Building Systems (800-200-6770), or in a pinch, I use a screwdriver to hold the nut flat while spinning it on with a pencil eraser. I tighten the nuts until the newel is secure and plumb, shimming as needed with

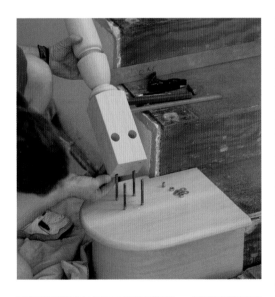

Hanger bolts provide a strong base for the starting newel. Four hanger bolts driven into the starting step will provide firm support for the starting newel. A specialty wrench (note rubber band) is used to start and tighten the nuts.

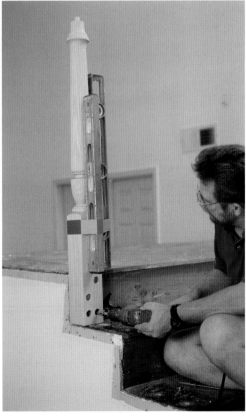

Landing newel is bolted to both riser and tread. Unlike the starting newel, the landing newel is bolted in two directions. The lower left hole provides access to the vertical hanger bolt; lag bolts are driven horizontally into the other three holes.

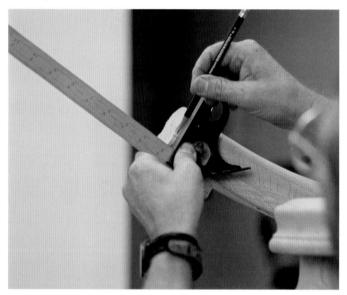

Taut string establishes the angle and height of the railing. The author fastens a stringline with tape between two fittings to check the position of the railing and to determine the proper angle at which to cut the fittings

Using a combination square to transfer the railing angle to the fitting. The author aligns the square's blade with the string (top) and marks the tangent on the fitting. Maintaining the blade angle (bottom), he flips the head over and draws the correct angle.

thin oak wedges. I try to use only one size of bolt throughout a job so that I need only one size washer, nut, wrench, and socket. After the handrail is installed, I go back and glue 1-in.-dia. plugs into the access holes.

Landing newels are the transition for railings between stairs and landings. I bolt them horizontally to the riser and vertically to the tread. I run one 5/16-in. hanger bolt into the step below that pulls the newel down; three 5/16-in. lag bolts pin it into the riser (see the bottom right photo on p. 243). After drilling the post and driving the hanger bolt into

the tread, I seat the post onto the bolt, apply glue, and screw the lag bolts into the riser. Two levels taped to the post speed up the adjustment process.

Using String to Ease Layout

There are two basic types of railings: post-to-post railings, which are cut to fit between newel posts; and over-the-post railings, which run uninterrupted over newel-post

Shop-made jig makes cutting odd-shaped fittings an easier and safer task. Fixed to the saw table by a bar clamp, the jig secures the fitting at the desired angle. Adjustment can be made by loosening the toggle clamp.

tops and use fittings and easings to change railing height and direction. Although there are many of these fittings, I installed two in the over-the-post railing featured here.

With the newels installed, I place the fittings, in this case a volute at the bottom and a gooseneck (see the photo on p. 241) on the landing, on the newel pegs. I drill and counterbore a hole into the top of the fitting for a screw, which I hide with a plug when the railing is done. I make sure each fitting is level and straight.

Next, I run string between the easings to mark the bottom of the handrail (see the left photo on the facing page). I center the string on the fittings, using masking tape to keep

it taut. I first see if the volute and gooseneck are rotated correctly by lining up the string with my square. I then check that the string is parallel with the stair and that it's at the corresponding height shown on the drawing.

I align the blade of the square parallel with the string under the fitting (see the top right photo on the facing page). This helps me to locate the tangent where the handrail and fitting meet. I then flip the square, keeping the blade parallel to the string, and mark the angle of the cut on the fitting (see the bottom right photo on the facing page). After cutting the fittings (see the photos above), I reattach them to the newels and

Locating Rail Bolts

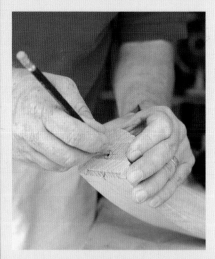

A thin slice of railing makes a good template for marking bolt holes. Using a template to mark bolt holes on mating pieces of railing will make a better match and save layout time.

Clamp the rails and fittings when drilling. A portable clamping bench is handy for keeping the railing parts steady while drilling. Both access and hanger-bolt holes should be as straight as possible to avoid misalignment of the parts.

Hanger bolts can be adjusted with a piece of hollow tubing or pipe. Occasionally, it's necessary to bend a hanger bolt carefully once it's in place. A foot-long piece of tubing exerts enough leverage and won't mar the threads.

run the string again to check the angles. Now I can measure and cut the straight railing sections. If an angle is off a bit, I can compensate by cutting a matching angle on the railing.

When all the rail parts are cut, I mark hanger-bolt locations on the ends of the matching parts with a slice of railing I use as a template (see the top left photo). After clamping the railing, I drill bolt holes and access holes in the railing (see the middle left photo), and drill the bolt pilot hole in the easing. If the bolts need to be tweaked, I carefully bend them with a foot-long piece of ¾-in. tubing, which won't mar the thread (see the bottom left photo). After bolting a section together, I check to see that the assembly fits on the respective newels. If it doesn't need fine-tuning, I screw the fittings to the posts, loosen one fitting at a time, squirt glue into the connection (see the top photo on the facing page) and retighten the fitting. Because it's rare that the rail and fitting profiles match exactly, once the glue is set I fair the two mating surfaces together with an assortment of tools that includes a 5-in. orbital disk sander, rasps, files, gouges and sandpaper. Fingertips will play over this joint, so it should be imperceptible to the touch.

Installing the Balusters

The balusters are fun to set because their installation signals that your hard work is nearly done. From the layout, I know I need to locate two baluster positions on each tread.

I align the forward face of the first baluster flush with the riser below and mark its location on the tread. After I mark the same spot on the tread above, I measure the distance from one mark to the other (see the bottom photo on the facing page), divide this measurement in half, and mark the middle baluster's position on the tread. When referenced from the outside of the

Glue strengthens the junction of rail and fitting. To make sure that the joint will not separate, the author spreads the joint apart with a screwdriver and squeezes PVA glue into the opening. The glue is dyed to match the wood.

For consistent baluster spacing, locate the front balusters and divide. After establishing the locations of the balusters at the fronts of the treads, the author measures the distance between the two points. The halfway point becomes the location of the second baluster.

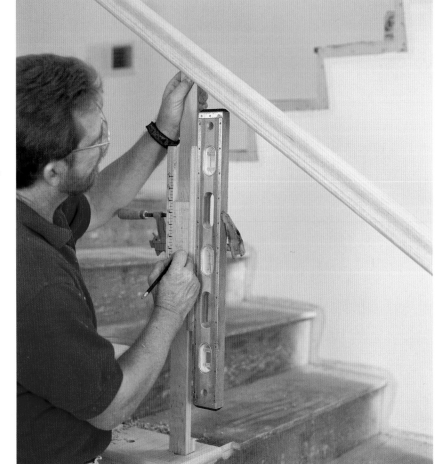

Sliding jig makes measuring balusters easy. Clamped to a level, this shop-built ruler is adjustable in height and quickly registers the height and handrail position of square-top balusters.

Fillet piece locks the balusters to the rail. After the balusters are pinned in place, thin fillet stock is cut, glued and nailed to the rail's underside, strengthening the entire assembly.

tread, these lines mark the locations of the baluster pin holes (these balusters have dowels turned on their lower ends). Again, I rely on the bubble level for drilling plumb holes.

I use another shop-built device to determine baluster lengths (see the photo above). Once I've cut the balusters, I glue the pins into the treads and nail the baluster tops to the underside of the railing. Thin pieces of fillet stock are cut individually and nailed between the balusters to lock them in place (see the photo at left).

I lay out balusters on a balcony or landing by first measuring in an equal distance from each newel. Using dividers, I find a baluster spacing that comes close to the spacing on the stairs. I always check these marks carefully before I drill any holes.

Lon Schleining is a stairbuilder in Long Beach, California, and also teaches woodworking at Cerritos College in Norwalk, California.

A Quick Way to Build a Squeak-Free Stair

■ BY ALAN FERGUSON

On a recent job, I overheard one of the young guys hanging cabinets in the kitchen say to his buddy, "Hey, who's the old geezer sitting on the stairs?" To my utter dismay, I quickly realized that there was only one set of stairs in the house, and the old geezer they were talking about was me. But thinking back on it, I have to admit that they were right. I've been building custom stairs for twenty-odd years, doing my own millwork and specialized handrails. As the years slipped by, I guess I have become that old man on the stairs.

But in the process of becoming this old stair geezer, I've learned a thing or two about building stairs, and I've reached a couple of conclusions. My first conclusion was that in order for a stairway to remain level, plumb,

An inexpensive stairway doesn't have to look cheap. Tread and riser end caps create a border for carpeting and a place of attachment for the newels and balusters. With shop techniques, even this economical stair can be built so that it stays level and squeak-free.

and squeak-free for a lifetime, it should be built under controlled conditions, such as in a shop.

With large stationary shop machinery, I can produce more accurate components that fit together to a T. Building the stair in a shop also means that I do all of my measurements and layouts in the friendly light and at the comfortable height of my workbench. Here, I can see my work properly to ensure accuracy. Too often on job sites, I end up working on my hands and knees on the floor. In the shop I always know there's an extra clamp nearby if needed, and there's no skimping or making do with less as I might be tempted to do on a job site.

On most sites, the handrail installer isn't part of the framing crew, so he's usually not around when the stairs are being built. But when I'm the guy who builds the stair and installs the handrail, I know where not to put fasteners and where to beef up the stair for extra strength, such as adding support for future newel posts. Another reason for not building a stair on site is that I always

feel I'm in the way of the other trades coming and going through the house, trying to get their work done.

One final plus to building a staircase in the shop is that I get to truck it to the site when it's finished. There's nothing more impressive than arriving at a site with a stair built as perfectly as a piano, carrying it into the house and nonchalantly slipping it into place. That's a sight that makes a hero out of any humble contractor in the eyes of a customer.

However, in spite of these advantages, most contractors in my neck of the woods like to build their own straight-run stairs on site. A guy like me could get mighty hungry waiting for a client or contractor to knock on the shop door to order a set of stairs. But I haven't been in this business all this time without learning to compromise.

So what about using shop techniques on site? That way, I have the advantage and convenience of building a plumb and level stair myself even if I can't have perfectly controlled conditions. My straight stair de-

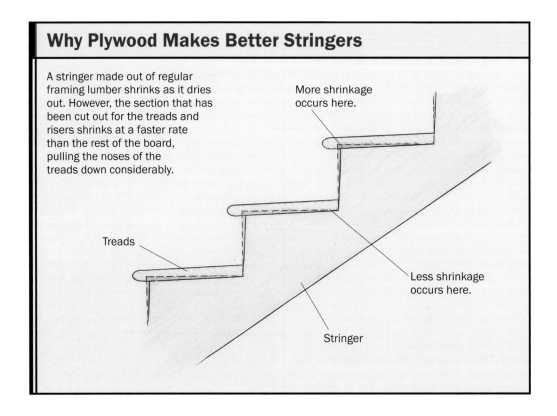

Why Plywood Makes Better Stringers

A stringer made out of regular framing lumber shrinks as it dries out. However, the section that has been cut out for the treads and risers shrinks at a faster rate than the rest of the board, pulling the noses of the treads down considerably.

More shrinkage occurs here.

Treads

Less shrinkage occurs here.

Stringer

sign will give you an affordable stair that can be built either in the shop or on the job site for permanently level, squeak-free perfection just by following a few simple shop techniques.

Plywood Stringers Eliminate Sloping Treads

Something else I've realized from my years of building stairs is that stair stringers, or carriages, should be made of a dimensionally stable material, such as plywood, instead of 2x lumber or, heaven forbid, green framing lumber. Building stairs on a site where moisture levels change radically intensifies wood movement and shrinkage. Stairs made with plywood stringers are more likely to remain level and plumb as the house heats up and experiences drastic moisture swings from the curing of concrete and drywall compound.

Stringers cut from standard 2x stock will shrink unevenly as moisture levels in the interior of the house stabilize. These stringers shrink less at the inside corners of the treads and risers (the narrowest points of the stringer), and more at the outside corners where the stringer is the widest. This differential shrinking pulls the nose of the tread noticeably downward.

Stairs with sloping treads are dangerous, and as a handrail installer, I shudder at the thought of sloping treads. Making good baluster cuts to fit sloping treads is difficult enough, but cutting mitered returns for these treads and risers with any precision is practically impossible. Stringers cut from plywood don't suffer the same shrinkage problems, which eliminates sloping treads.

Materials Are Chosen to Fit the Budget

My stair design is versatile and can be adapted to almost any finishing style and budget simply by selecting different materials. The stair in this chapter was an economical solution for first-time homeowners struggling to stay within budget. Although they would have preferred hardwood treads, they opted for carpeted treads with exposed hardwood tread end caps (see the photo on p. 249). I cut the treads from off-the-shelf 16-ft. lengths of 1-in. OSB tread stock that come $11\frac{3}{4}$ in. wide with one bullnose edge.

Like the treads, the stringers can receive a variety of treatments, such as paint, veneer, or addition of a stylish trim. The stringers I used on this stair were to be covered with drywall, then painted, so I chose $\frac{3}{4}$-in. poplar exterior-grade plywood; however, any high-quality exterior plywood would have worked.

The risers were also to be carpeted, so I cut them out of $\frac{1}{2}$-in. fir plywood. I always use kiln-dried 2x2 cleats for the riser-to-tread connections. I'm a stickler about using good-quality cleats to guarantee the best connections, so I buy baluster manufacturers' blanks that could not be used because of minor blemishes. Baluster companies are usually happy to sell me bundles of these seconds, called "sweet ones" in the industry, at a reduced price.

Two Layers of Plywood Form the Stringers

My first task was ripping the plywood for the stringers into $11\frac{1}{2}$-in. widths. Using 8-ft. and 4-ft. lengths for each layer, I laminated two layers of plywood together to form stringer stock $1\frac{1}{2}$ in. thick and 12 ft. long (see the photo on p. 252), which I had estimated to be just long enough for this stair. Because the house was already weathertight and because there was no danger of the staircase getting rained on, I was able to

Laminated stringers won't shrink or warp. Two layers of ¾-in. plywood are glued together for stringer stock that will remain stable and true for the life of the stair.

use one-part, shop-grade white glue with a 20-minute setup time. I staggered the butt joints of the short and long pieces between the layers and glued the butts with construction adhesive. If rain or moisture had been a consideration, I would have used an exterior waterproof glue.

I brought a big portable workbench to the site to give me a flat working surface for efficient clamping and gluing, but I could have easily set up a temporary workbench on site. Most guys would have skipped this step and done their glue up on the plywood deck, but the glue up is much easier to control on a good workbench.

For fast glue application, I use a fuzzy 3-in. paint roller with the glue in a pail. After spreading glue on both layers of plywood, I mated the top and bottom layers together and clamped them every 12 in. using C-clamps.

I always double-check that my clamps are all tight, which means there will be no glue voids. While the glue was setting up on my stringer stock, I cut the tread and riser stock.

To determine the depth of my treads, I first had to calculate the width of the nosing. The end caps I chose for the treads came with 1¾-in. nosing, but the tread stock was to be covered with ½-in.-thick carpet that my customer had selected. To make the overhang of the carpeted tread flush with the end cap, I added 1¼ in. (1¾ in. minus ½ in.) to my unit run and ripped the OSB tread stock to that width. Ripping the riser stock to my unit rise was straightforward.

Leaving Out the Squeaks

When the glue had set, I unclamped the stringer stock and began my layout. Being careful to allow for pencil-lead thickness for each tread and riser, I laid out the stringer with my framing square and stair gauges (see the photo on the facing page). Yes, I know it's a pain, but you'll thank yourself for making that extra trip out to your truck to get your stair gauges. Without them, it's impossible to duplicate the exact measurements from one tread/riser combination to the next. Just about or close enough at this point won't yield a precision stair.

For most stairs total run and width measurements come from the framed opening in the floor. Rise measurements are normally taken from finished floor to finished floor. However, these plans called for a landing with one step down at the start of the run. This detail meant that I needed to take my total-rise measurement from the top of the landing deck to the top of the second-floor deck. (For more on calculating rise and run for stairways, see "Cutting Out Basic Stairs" on p. 226).

The completed layout included the bottom foot cut where the stairs sit on the landing as well as the joist cut at the top of the stringer. The joist cut lets the stringer run up to the ceiling rather than end in an abrupt drop from the deck at the top of the stairs. I believe that attention to visual details such as this one is what distinguishes a craftsman's work.

Having determined the exact size of the bottom foot on the stringer, I gave the framer precise measurements for the height of the landing, which he began immediately so that it would be ready when I finished my stair. When the framer built the landing, I made sure that he glued and screwed every connection, which is the greatest factor in making squeak-free stairs.

I always add plywood backing to the top riser where the stairs attach to the upper deck. This extra layer of plywood helps to stiffen the stairway during installation and provides extra reinforcement under the nosing of the top landing. Because this small house was built with 8-in. rather than 10-in. or 12-in. floor joists, I cut a 2-in.-deep mortise in the stringer at the inside corner of my joist cut that allowed me to use a full 10-in. width for my extra plywood backing.

Each tread/riser layout is duplicated exactly. Stair gauges clamped to the edge of a framing square allow the layout to be repeated precisely for each step.

What Makes Stairs Squeak?

Squeaking stair treads start with the shrinking and warping of wood as it dries, which allows the nails to work loose. What you hear when you step on a squeaky tread is the sound of the wood moving against a nail. All framing nails, even galvanized, tend to work loose, whereas screws seem to stay put. My theory is that the glue fills any voids and that the screws hold everything together. The result is silent stairs.

Simple Jigs Speed Up Cutting and Assembly

Next I make a jig from MDF scraps to act as a fence for my circular saw to follow when I make riser and tread cuts in the stringer (see the photo on p. 254). The jig has a fence that registers against the bottom of the stringer, and I line up the jig with my layout, using a wooden block whose width is the same as the distance from the edge of the saw's foot to the sawblade.

Using a jig to make the tread and riser cuts is another shop technique most carpenters working on site would probably skip,

TIP

Add plywood backing to the top riser where the stairs attach to the upper deck. This extra layer of plywood helps to stiffen the stairway during installation and provides extra reinforcement under the nosing of the top landing.

thinking it too much trouble. But using this jig as a guide, I eliminate problems and imperfections that result from freehand cutting with a circular saw. My cuts with the jig are perfect, ensuring cabinet-quality joinery for the finished stair.

Because the circular saw doesn't reach all the way into each corner, I finish each cut with a jigsaw and then clean up the inside corner with a wide chisel. When the first stringer is finished, I clamp it securely to the second one and trace it to ensure an exact match. I cut the second stringer the same way as the first.

Now, with both stringers cut, I make a simple assembly jig that screws to two sawhorses (see the top photo on the facing page). This jig holds the stringers firmly and perfectly parallel at waist height. Here again, most of the guys I see on site either assemble the stairs on the floor, working on their hands and knees, or they assemble the stairs in place, struggling to fasten the treads and risers while keeping the stringers aligned properly.

Next I square one stringer to the other in the assembly jig using witness marks and my framing square (see the bottom photo on the facing page). The witness marks are made by placing the stringers flat against each other and making a pencil mark across both of them at some point. Then, with the stringers in the jig, I line up both pencil marks using a framing square and an extended straightedge that leaves my stringers perfectly square to each other in the jig.

A note of caution here about stair width. When determining stair width, be aware of any compensations you might need to make for drywall or to make the stairs fit in the rough opening. This stair was being installed along a wall, so I allowed for a 1-in. space for drywall installation. I also had to ensure that the finished stair would fit the rough opening, so I subtracted an additional ¼ in. when figuring the correct distance

MDF jig for perfect sawcuts. The table of the circular saw registers against the edge of a simple jig made from MDF to eliminate freehand sawing. The cuts are finished with a jigsaw and fine-tuned with a chisel.

Assembly Jig Serves as a Workstation

Assembly jig holds the stringers parallel and square. A simple jig made from strips of plywood holds the stringers at the proper width while the author is attaching treads and risers.

My assembly jig took a few minutes to set up, and it gave me a workstation at a comfortable height where I could keep things square and level. The assembly jig is made of two 8-ft.-long platforms mounted between two sawhorses. Each platform consists of a 3-in.-wide plywood cleat screwed to a 10-in.-wide base piece. These platforms are screwed to the sawhorses so that when the stringers are standing up against the cleats they are at the desired stair width. Some simple spreaders between the stringers hold the stringers firmly in place against the cleats.

between the stringers. This extra space gave me ample clearance to slip the stair easily into place in the rough opening of the second-floor deck.

Staples Hold the Stair Together Initially

After the treads and risers were cut to length, the stair was ready for assembly. As I mentioned before, I always glue and screw all surfaces on all of my connections to ensure a squeak-free stair. Initially, however, I assemble the stair with a few staples, just enough to hold the elements together while the glue dries. At that point, I go back and add screws to every connection.

I start my assembly with the bottom riser. When it is glued and stapled in place, I set in one of my 2x2 "sweet" cleats behind the

Witness marks help align the stringers in the jig. Lines called witness marks are made with the stringers laid side by side. Then a square and a straightedge align one stringer perfectly with the other.

Risers go on before the treads. For each step the riser is glued and stapled to the stringers ahead of the tread. Staples driven through the backs of the risers into the treads hold them together until the glue dries.

top edge to give me sufficient material to screw the tread into. Before putting in the first tread, I install the second riser and cleat (see the photo above). Then I glue the first tread in place and fasten it with one staple in each stringer and a couple of staples in the riser cleat. I also drive a couple of staples through the plywood riser into the tread, which has to be done from the backside of the riser. Following the same process, I staple and glue the rest of the risers and treads in place except the top tread and riser.

At this point I preassemble the top riser and nosing together with the piece of backing plywood that fits into the stringer mortises I made earlier (see the left photo on the facing page). I hold this assembly in place while I lay in the last tread and staple it. This special beefed-up riser is one of the trademarks of my stairs. It not only stiffens the top riser for handling (and transporting), but it comes in handy when I'm setting the stairs in place.

Before removing the stair from the jig, I screw the treads to both the stringers and the riser cleats using 2-in. coarse-thread floor

Screws keep out the squeaks. After the glue has set, screws are driven to fortify every joint and ensure a squeak-free stair.

screws with square-drive heads (see the right photo above). I use these screws because they're beefier than drywall screws, so normally I don't have to predrill. In fact, the only predrilling I do is when I screw the top nosing to the top riser and backer. I finish the top assembly with a few screws through the backer into the stringer at the joist cut. I also glue and staple a 1-in. spacer strip (see the photo on p. 258) to the edge of the wall-side stringer.

Now I'm ready to flip the stair up on its side to screw through the back of each riser into each tread on approximately 6-in.

centers. As a final touch, I add a 2x2 cleat under the top tread against the ½-in. plywood backer where it extends down past the top riser. And the stair is ready for installation, barely six hours from when I started.

The Top Nosing Acts as a Handle

Ordinarily, a simple straight stair such as this one could be installed the same day. But those two young carpenters in the kitchen got so tuckered out just watching me whip up these stairs that I decided to postpone

A spacer strip allows room for drywall. A 1-in. strip stapled to the stringer will keep the stringer away from the wall studs so that drywall can be slipped in behind the stringer after the stairs are installed.

the installation until the next day when they were fresh.

The following day, I arrived to find my two young friends and the framer waiting to help with the installation. After double-checking the landing dimensions, we lifted the stair into place for a dry-fit. Now here's where that top-riser/nosing assembly comes in handy. I got on the top deck, and with three guys lifting the stair, I reached out, grabbed the top of the stair by the nosing and guided the whole thing into place.

The stair went in beautifully, so we took it back out to prepare for the final fit. I ap-

plied construction adhesive liberally to the joist header and under both stringer feet on the landing; then we installed the stair for good. My two crucial checkpoints were making sure that the 1-in. spacer strip was snug against the wall and that the top nosing was flush with the deck.

When everything was lined up properly, I quickly drove a nail through the top riser into the joist header to hold the stair in place temporarily. The guys down below could now let go and breathe easily while I took my time securing the stair through the beefed-up riser with 2-in. screws. Then

Ready for traffic. With the stair screwed in place and the step to the landing installed, the stair is ready for use by the construction crew. End caps and balusters will be added later during the trim stage. (Note: Insulation scraps made a great sound barrier for the wall behind the stairs.)

I moved to the bottom and angle-screwed through the stringer feet into the landing. Finally, I screwed the wall-side stringer to the studs with long screws.

In my spare time the day before, I built the single step leading up to the landing by gluing and screwing a box together using offcuts from the OSB treads, and I secured the step in place with screws and construction adhesive. The wooden end caps would be installed when I came back to install the handrail. And voilà, much to the relief of the other subcontractors and to the joy of my customers, the stair was ready for use with a humble price tag but a picture of close-to-cabinet-quality precision and squeak-free perfection.

Alan Ferguson currently works as a kitchen and bath designer on Vancouver Island after a 27-year career as a designer and builder of custom furniture and architectural millwork. He resides in Qualicum Beach, British Columbia, Canada.

A Built-In Hardwood Hutch

■ BY STEPHEN WINCHESTER

I love an old house. Working on one makes me appreciate the skill of the carpenters who came before me. It's amazing to see the level of craftsmanship the old-timers attained using only hand tools—especially in their trimwork. I recently renovated an early 1800s farmhouse in New Hampshire that had some beautiful chestnut trim. I got the chance to match this woodwork when I added a family room with a built-in hutch.

I made the new family room by removing a wall between two small rooms. There was a closet in each room, one on both sides of the wall, and when the wall came down, the closet area was a natural location for the built-in hutch. Built-ins ought to look good and last a long time, so this hutch was built of solid hardwood and designed to accommodate wood movement (see the drawings on pp. 262–263). But before I started building, I straightened and leveled the closet area.

Roughing In the Hutch

New studs on the left and right made the sidewalls plumb and straight, but there wasn't room on the back wall for new studs. So I straightened the back wall with shims and 1x3 strapping (bottom photo, facing page). At the bottom I tacked a 1x3 across the old wall and into the old studs. Placing a straightedge on the 1x3, I tucked some shims behind the low spots to bring them out to the straightedge. Next I tacked a 1x3 to the top, again shimming it straight. Then I tacked on more horizontal 1x3s 16 in. o.c. Moving from left to right, I held the straightedge vertically, against the top and bottom strapping, and shimmed the intermediate strapping out to the straightedge. The wall was straight when all the pieces of strapping were even with each other.

The hutch rests on a 2x4 base; I installed it level by shimming the low end and nailing it to the new 2x4 walls on each side. With the new level base, I didn't have to scribe the cabinet sides and back to the floor, which had a big hump in it.

Ash matches. A new ash hutch built into an old closet looks like the chestnut woodwork of the original room. The hutch was finished with two coats of Minwax® Polyshades®—half maple and half walnut—followed by a slightly thinned top coat.

The opening. New studs frame the sides, but the back wall of this former closet was straightened with 1x3s wand shims.

Chestnut Substitute

Chestnut was once used for almost everything in a house, from sheathing to door and window frames to trim. But during the first part of this century, a blight wiped out almost every American chestnut tree. Today, you can get salvaged chestnut from old buildings or get it resawn from beams or sheathing, but it's expensive. I chose white ash instead, which has about the same grain pattern and texture as the chestnut woodwork on this job. But ash is hard, so it's more difficult to work than chestnut.

Gluing Up Wide Boards

The cabinet floor and the counter were glued up out of several boards, as were the wide shelves for the bottom cabinet. To joint and join the boards in one step, I used a glue-joint cutter in the shaper. (Jointing is the process of straightening a board's edge or face and is typically done with a jointing plane or with an electronic jointer. Joining is the process of connecting two boards.) The glue-joint cutter makes edges that look something like shallow finger joints (bottom detail drawing, facing page). These edges align the boards and provide a larger gluing surface than simple square edges do. Glue-joint bits are also available for use in router tables.

First I lined up the boards so that their grain matched, and I marked them so that they wouldn't get mixed up during the glue-jointing operation. I used numbers—1s on the first two adjoining edges, 2s on the next two, and so on.

I don't have a wide planer, so I had to flatten the glued-up boards with a belt sander. With a 60-grit belt, I sanded across the grain first, then with the grain. Then I used a 100-grit belt and finished with a 120-grit belt. The countertop, the most visible of these wide boards, was finished using 180-grit paper on a random-orbit sander.

Spline-and-Groove Wainscot

One of the original small rooms had beaded wainscot all the way around, so I decided to use beaded wainscot inside the hutch. To make the wainscot, I ripped ash boards on the table saw into random widths, from 5¼ in. to 3¼ in.

To join the pieces, I used a spline-and-groove joint rather than a tongue-and-groove joint (top detail drawing, facing page). First I jointed the edges of each board. To make the groove, I used a ¼-in. straight

Putting the Carcase Together

Countertop notched to accept full-length wainscot.

Wainscot divider is biscuited in place.

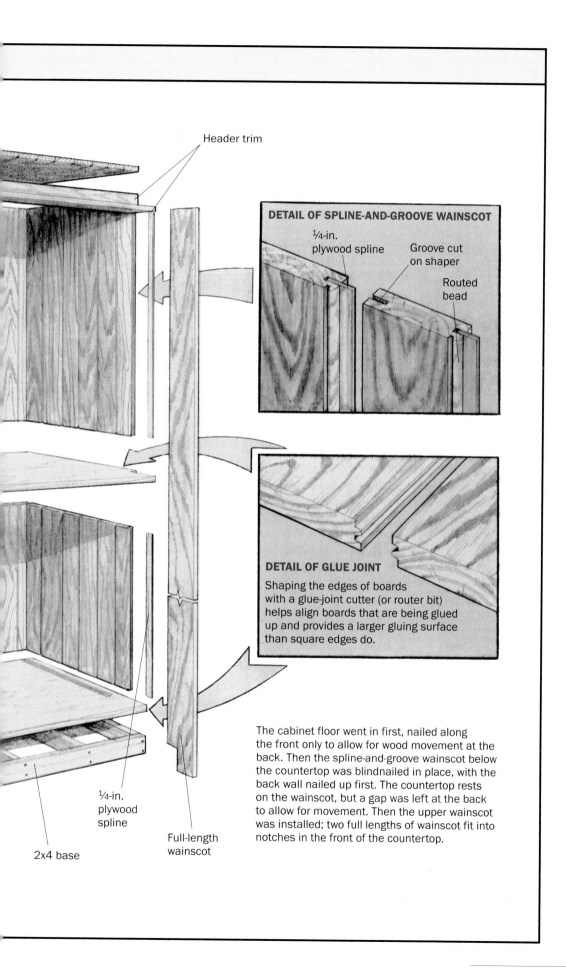

Header trim

DETAIL OF SPLINE-AND-GROOVE WAINSCOT

¼-in.
plywood spline

Groove cut
on shaper

Routed
bead

DETAIL OF GLUE JOINT

Shaping the edges of boards
with a glue-joint cutter (or router bit)
helps align boards that are being glued
up and provides a larger gluing surface
than square edges do.

¼-in.
plywood
spline

2x4 base

Full-length
wainscot

The cabinet floor went in first, nailed along
the front only to allow for wood movement at the
back. Then the spline-and-groove wainscot below
the countertop was blindnailed in place, with the
back wall nailed up first. The countertop rests
on the wainscot, but a gap was left at the back
to allow for movement. Then the upper wainscot
was installed; two full lengths of wainscot fit into
notches in the front of the countertop.

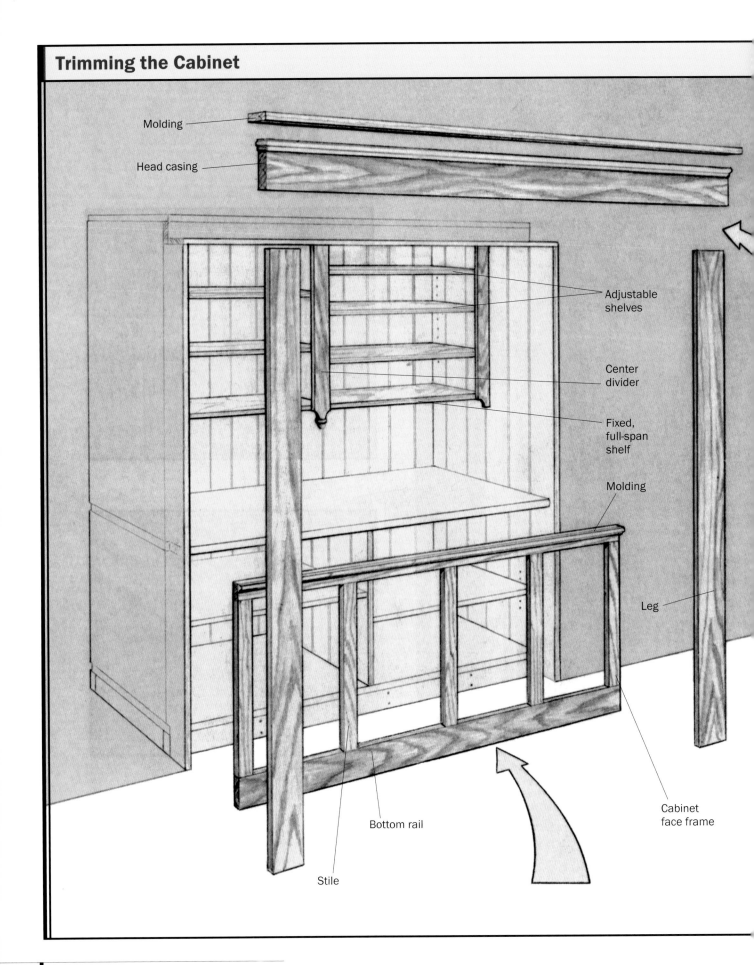

Molding

Head casing

Adjustable shelves

Center divider

Fixed, full-span shelf

Molding

Leg

Cabinet face frame

Bottom rail

Stile

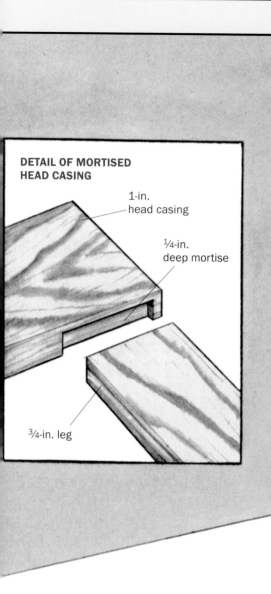

DETAIL OF MORTISED HEAD CASING

1-in. head casing

¼-in. deep mortise

¾-in. leg

The cabinet face frame was assembled, then glued and screwed in place. The head casing was mortised to fit over the legs to keep the joints from opening up. A fixed, full-span shelf above the countertop supports a divider and adjustable upper shelves.

cutter on the shaper, but a ¼-in. slotting cutter in a hand-held router or a dado-blade assembly in the tablesaw would work, too.

I centered the ½-in.-deep groove on the edge of the board. The 1^{15}/$_{16}$-in. splines were ripped from ¼-in. plywood. I didn't use biscuit joinery because, when wainscot shrinks, gaps appear between the biscuits. A full spline looks like a solid tongue.

Using a beading bit, I beaded one edge of each board to match the original chestnut wainscot.

Installing the Wainscot

I installed the floor of the cabinet first, flush to the front of the 2x4 base. I nailed the floor at the front only and left a ⅜-in. space at the back to allow for wood expansion.

I put up the wainscot for the bottom half of the hutch by blind-nailing through the splines and into the walls as I would any T&G material. I didn't glue the splines because each piece of wainscot should expand and contract independently. This wainscot rests directly on the cabinet floor; if the floor butted into the wainscot, a seam would open. I avoided visible seams in the corners by putting up the back wall first and then butting the sidewalls into it. And I allowed for expansion by installing the first board on the back wall ⅜ in. from the corner.

I also made a wainscot divider for the bottom cabinet. It was biscuited to both the floor and the underside of the counter. I used just a dab of glue in each biscuit slot to prevent any unnecessary glue squeeze-out.

The counter sits on the wainscot. Before I installed the counter, I notched its two front edges, which would allow an entire length of wainscot at the front of each sidewall. Along the sides, the counter is nailed into the wainscot so that it stays put, but to allow for expansion and contraction, the back edge of the counter isn't nailed.

Now I was ready to put the wainscot in the top of the hutch. I set the wainscot on the counter and blind-nailed it through the splines into the walls. Putting the wainscot up in two sections, bottom and top, eliminated the wood-shrinkage gaps that would have resulted from running the wainscot from floor to ceiling and butting the counter into the wainscot. With the front edges of the counter notched, I installed the front pieces of wainscot on each sidewall. Because the unit is recessed into the opening, I wanted a full length of wainscot from floor to header with no seam.

Pocket-Screw Joinery

In my shop, stiles and rails for the face frame were cut to width but not to length. Stiles and rails are the vertical and horizontal frame pieces, respectively.

I assembled the face frames on site. I cut the stiles and the rails to length and clamped them to the cabinet to check the fit. After some slight trimming on a compound-miter saw perfected the face-frame joints, I laid the stiles and the rails on the bench and screwed them together.

The top rail was narrow enough to allow the stiles to be joined to it with screws driven straight through the edge. But the bottom rail of the cabinet was wide, and the intermediate stiles butted into it, so here I screwed the rail to the stiles through pocket holes. A pocket hole is a cut made on the face of a board that doesn't reach the board's edge.

There are several jigs on the market to make pocket holes—from simple guides for a hand-held drill to dedicated pocket-hole machines. I don't have any of them, so to make the pocket holes to assemble this face frame, I used a spade bit, starting the hole with the drill held vertically and tipping the drill back as I fed the drill bit in (see the top photo). The pocket hole ended at a mark 1½ in. from the edge of the rail. Then I drilled a pilot hole in the edge of the rail at an angle up through the pocket hole (see the middle photo). Finally, I squeezed a generous amount of glue between the stiles

Pocket-screw joinery. To attach the bottom rail to the stiles, a spade bit makes a pocket hole that's 1½ in. short of the rail's edge (top). A pilot hole is then drilled up through the edge to connect with the pocket hole (middle), and the boards are glued and screwed together (bottom).

and the rails, clamped them together and ran the screws in (see the bottom photo).

After the glue was dry, I sanded the joints flush and installed the face frame. I glued the bottom rail to the front edge of the cabinet floor and screwed the top rail to the underside of the countertop (see the drawing on pp. 264–265).

Mortised Head Casing

The trim, or casing, around the hutch was installed next. The ¾-in.-thick side pieces, or legs, went on first; I ran them ¼ in. long at the top. The 1-in.-thick top piece, or head, was mortised to fit over the legs (see the detail drawing on p. 265). You could think of this as being a mortise-and-tenon joint, with the legs being the tenons. I set the head on top of the legs and with a sharp pencil traced the outline of each leg onto the bottom edge of the head. Then I scored the marks with a sharp knife. Scoring makes for a cleaner mortise. I mortised these sections of the head a good ¼ in. deep with a hinge-mortising bit in my small router. Finally, I used a chisel to square the corners of the mortise. This joint practically guarantees lasting beauty: If the header shrinks, the joint still looks tight.

I wanted the molding under the front of the counter and at the top of the head casing to match the original molding at the top of the doors and the windows (see the photo below). This molding wasn't something I could have picked up at the lumberyard, and I couldn't find any cutters the right shape, so I combined two different shaper cutters to make the molding (see the drawing at right). The result was a perfect match.

Matching Molding

Bead

1. The author started with a length of 1¹⁄₁₆-in. by 1⅛-in. ash stock.

2. The first pass was with a ⅜-in. bead cutter in the shaper.

3. Below the bead, waste was ripped on the table saw ³⁄₃₂ in. deeper than the flat above the bead.

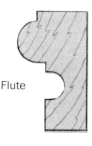

Flute

4. Then the shaper was used to cut a ¼-in. flute.

5. Another pass on the tablesaw removed ³⁄₃₂ in. of waste below the flute.

Cove

6. A ¼-in. flute cutter on the shaper cut a cove to complete the profile.

Making Doors

I made frame-and-panel doors for the cabinet at the bottom of the hutch.

The door stiles are 2 in., the top rail is 2½ in., and the bottom rail is 3½ in. After cutting the pieces to size, I used my shaper to mold the inside edges of the frame, cut the panel groove, and made the cope-and-stick joint between the stiles and the rails.

I assembled the frames dry to check the door size and to get the panel size. I allowed ⅛ in. on each side of the panel for expansion. The ash panels on these doors were raised (beveled around the edges) on the shaper, so I glued up the boards with square joints to make the wide panels. If I had used the glue-joint cutter, the glue-joint profile would have been visible when I shaped the raised edges.

To be sure everything fit, I dry fit the panel within the frame before gluing up. Then I glued the doors and clamped them.

I used a small amount of glue on the joints because the squeeze-out could glue the panel in place, and the panel should be free to expand and contract.

Stephen Winchester is a carpenter and woodworker in Center Barnstead, New Hampshire.

A tablesaw and a shaper were used to make ash molding (right) that matches the original chestnut trim (left).

Designing and Building an Entertainment Center

■ BY BRIAN WORMINGTON

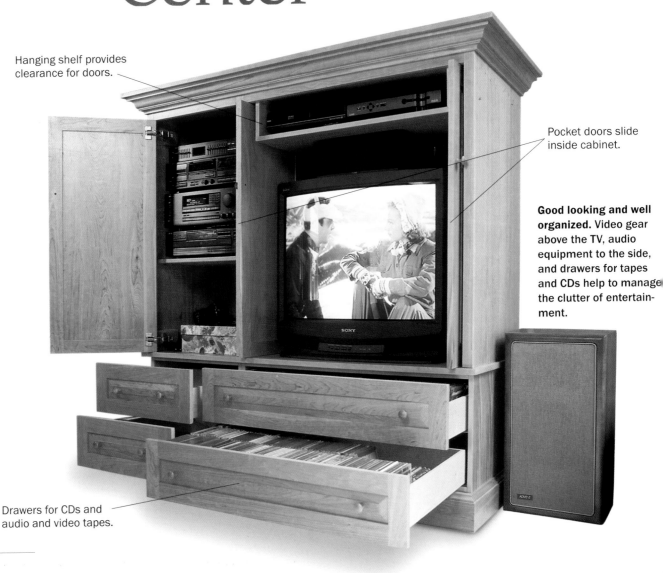

Hanging shelf provides clearance for doors.

Pocket doors slide inside cabinet.

Good looking and well organized. Video gear above the TV, audio equipment to the side, and drawers for tapes and CDs help to manage the clutter of entertainment.

Drawers for CDs and audio and video tapes.

The first extendable TV turntable I installed in an entertainment center was a surprise. This slick hardware, rated to hold 200 lb., seemed to be an engineering coup. I smiled as I helped my customer set his new 32-in. TV in place. But as he extended the turntable, I started to worry: The TV bobbed up and down alarmingly. The bobbing worried my customer, too.

He never again pulled the TV out from the cabinet—the image of his $900 TV on the end of a diving board was too disturbing.

I run a one-man shop where custom entertainment centers are my main business. Years of handling the whole process—designing, building, and installing—have taught me to avoid such gaffes as flexible turntables.

Most Entertainment Centers Revolve Around the TV

Designing an entertainment center begins with a visit to the client's home. I measure the space, see where the TV will be most easily viewed, and verify the presence of electrical and cable outlets. Sometimes there is a radiator or heat vent in the way, and the client needs to be reminded to call the appropriate contractor to move it.

I build freestanding and built-in units. My techniques apply equally to both types. Because I work alone, I build entertainment centers in modules (see the drawing on p. 270). One large cabinet would be difficult to handle, but I can move smaller modules by myself and join them together on the job.

The design of most entertainment centers revolves around housing a 27-in. or 32-in. TV (see the drawing above). Such large TVs are 2 ft. or more deep. Although the dimensions don't vary much among TV manufacturers, I always measure the set before designing the cabinet.

Depending on the final height of the TV shelf, I fill the space below it with two or

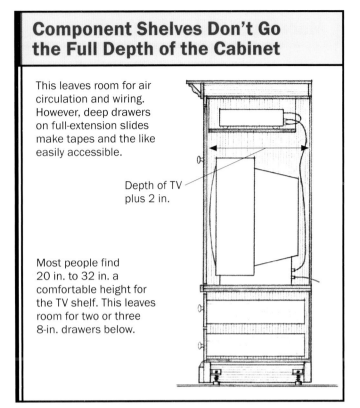

Component Shelves Don't Go the Full Depth of the Cabinet

This leaves room for air circulation and wiring. However, deep drawers on full-extension slides make tapes and the like easily accessible.

Depth of TV plus 2 in.

Most people find 20 in. to 32 in. a comfortable height for the TV shelf. This leaves room for two or three 8-in. drawers below.

Aluminum bars divide drawers. These drawer dividers are set up for videotapes, but the wood blocks where the bars rest can be flipped over. Their other side has slots to arrange the bars to organize CDs and audio cassettes.

three full-extension drawers to store CDs, audio tapes, videocassettes, and accessories.

CDs are 4⅞ in. wide, and audio tapes are 4¼-in. wide. I divide my standard 20¼-in.-deep drawers into 5-in. rows that accommodate either. Videotapes take up 7½ in., so I divide drawers for them into three rows. The front two rows are the full width of the tapes, while the back row holds several rows of four tapes lengthwise. Reversible blocks inside the drawers hold aluminum bars that divide tapes and CDs when set one way, and videocassettes when set the other way (see the photo above).

Modules for Easy Installation

Rather than assemble large units in the shop, the author transports easy-to-handle sections to the client's home for assembly.

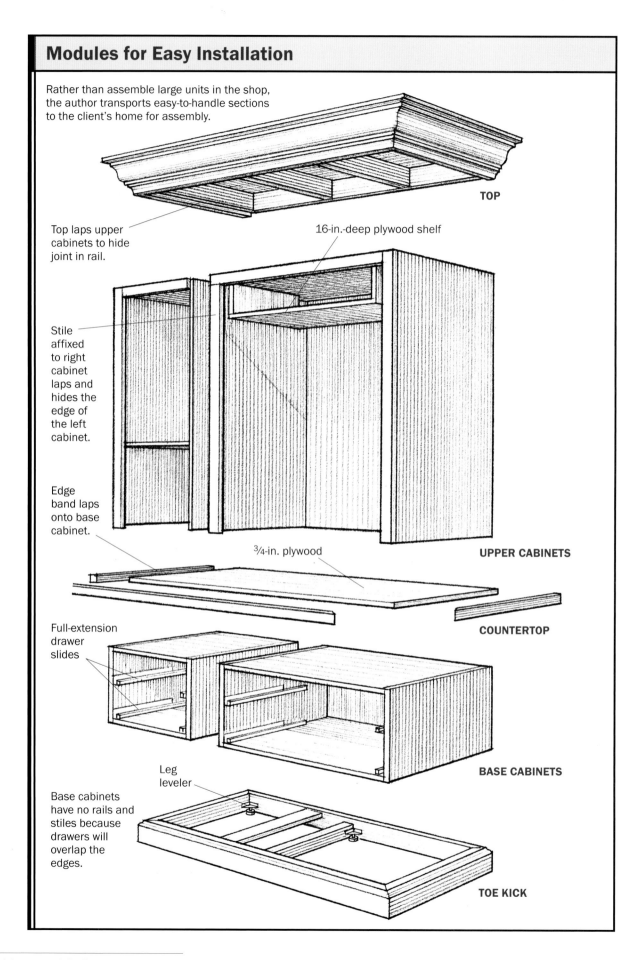

TOP

Top laps upper cabinets to hide joint in rail.

16-in.-deep plywood shelf

Stile affixed to right cabinet laps and hides the edge of the left cabinet.

Edge band laps onto base cabinet.

3/4-in. plywood

UPPER CABINETS

COUNTERTOP

Full-extension drawer slides

BASE CABINETS

Leg leveler

Base cabinets have no rails and stiles because drawers will overlap the edges.

TOE KICK

Biscuits and Pocket Screws Hold the Cabinet Together

I build these units from ¾-in. hardwood-veneer plywood. I butt-join the tops, bottoms, and sides, using #20 biscuits to align the parts and pocket screws to hold them together. To keep the case square, I rabbet the sides, top, and bottom to receive a ½-in. plywood back. I screw the back to the case with 1⅝-in. cabinet screws. The back is strong enough for the 2½-in. screws I use to attach the cabinet to the wall. I don't need an extra cleat. If necessary, the back can be removed easily to route electrical wires.

Hidden levelers ease installation. After assembling the entertainment center in the house, the author adjusts the levelers through holes that will be concealed by drawers.

Put the VCR and the TV in the Same Module

Once the location of the TV is decided, the arrangement of the other equipment more or less falls into place. I find that the best place for the VCR and the cable or satellite box is on a shelf above the TV. This placement simplifies wiring, and the TV and its peripherals can hide behind the same doors. That way, if the TV is on, the doors in front of the VCR are open. This is important because remote controls communicate with their components by line-of-sight signals. There are devices that sidestep this problem by relaying the signal. But they cost about $100, and the need for them is avoided by keeping the VCR with the TV.

If there are only a few components, it may be possible to stack the audio and the video gear together on a shelf above the TV. Most equipment is the same width, 17¼ in., and about 1 ft. deep, so it stacks well.

But few people trouble with an entertainment center to house only a few components. They either own or intend to buy a serious system. I often build three-door cabinets to house such systems. Two doors enclose a TV cabinet, and the remaining door closes on an audio-component cabinet about half the width of the TV cabinet.

Books Are Entertainment, Too

Often, clients want bookshelves as part of their entertainment centers. Bookshelves generally aren't as deep as shelves for audio or video components; a 12-in. shelf accommodates most books. I build plywood bookshelves in modular units no more than 32 in. wide, a dimension that minimizes plywood waste. It's also a maximum width that I'm comfortable with for a ¾-in. plywood shelf. Wider shelves sag more under the weight of books. Of course, it's possible to build wider shelves that don't sag by laminating several layers of plywood together. Another way to beef up a shelf is to replace the regular ½-in. by ¾-in. hardwood edge with 1x2 or 1x3 stock.

Keeping the Music Cool

A frequent concern my clients have is cooling their electronic equipment and TV. This concern is a holdover from the days of tube-based equipment, which generated considerable heat. Modern solid-state components don't generate much heat. Some manufac-

Adjustable Shelves That Stabilize a Tall Bookcase

To accommodate clients who are interested in showing off extensive libraries, I build bookcases. To stiffen wide or tall bookcases, it's sometimes necessary to fix a centrally located shelf in place. Rather than permanently install such shelves, I preserve some adjustability by affixing the shelf with Titus knock-down connectors. These connectors have studs that thread into shelf-peg holes. Connectors that are set into holes bored in the shelf bottom capture the studs. A reverse turn of the screwdriver drives the parts tightly together, bracing the cabinet and making it stable for plenty of books.

Case sides are bored for shelf pegs on 32mm centers. The studs for these connectors screw into the peg holes. They can be moved to different holes to improve the shelf layout, something that's impossible with fixed shelves.

turers recommend about 7 in. of airspace above the components and a few inches to the sides and back.

Sometimes, however, I encounter tube-based equipment and large power amplifiers. They need air movement; the manufacturer can tell you how much. In these cases, I drill ventilation holes in the top or in the back of the cabinet and ventilate with whisper fans. These small fans are commonly used to cool computers. I buy them for about $12 each from JDR Microdevices®.

TVs don't need much ventilation. The picture tube is taller than the vents in the TV's back, so there is always airspace above. And the cabinet doors are open when the TV is in use, so airflow is constant.

Provide Access to Equipment Backs

I encourage my clients to stack their components—except for heat-generating tube equipment—atop each other. I provide a pull-out shelf on a full-extension drawer slide to stack the components on. With the shelf extended, there is easy access to rear-panel cable connections.

An exception is when the client has a record turntable or carousel-type CD player that loads from the top. I provide individual pull-out shelves for these items.

Hiding the TV

Many of my clients want to be able to close the TV behind doors. I think it's to hide the fact that they actually watch TV. Sometimes, regular hinged doors won't work. There may not be room to open them all the way, or the client simply may not like the look of open doors. This is why pocket doors that open 90 degrees and slide into the cabinet alongside the TV are a popular option (see the top left photo on the facing page). The cabinet width must be increased by 2 in. on a side, or 4 in. total, to accommodate the pocket-door hardware.

Pocket doors hide that embarrassing TV when guests visit. Then they slide unobtrusively to the sides of the TV as the set returns to regular use.

Shelf hangs from cabinet top. Angle iron on the shelf's side fits into rabbeted blocks above the pocket-door slides.

Shelf doubles as a doorstep. Stops on the countertop might interfere with sliding in the TV.

Sources

Accuride International
12311 Shoemaker Ave.
Santa Fe Springs, CA 90670
562-903-0200
www.accuride.com

Julius Blum & Co.
PO Box 816
Carlstadt, NJ 07072
800-526-6293
www.juliusblum.com

JDR Microdevices
1850 S. 10th St.
San Jose, CA 95112
800-538-5000
www.jdr.com

Knape & Vogt Manufacturing Co.
2700 Oak
Industrial Dr. NE
Grand Rapids, MI 49505
616-459-3311
www.kv.com
KV 1385

Titus
22020 72nd Ave. S.
Kent, WA 98032
800-762-1616
www.titusint.com

Because pocket doors slide into the cabinet, they are necessarily smaller than their opening. Therefore, they can't overlay the cabinet face and must be inset. For symmetry, I often inset the other doors on the unit as well, using Blum® European-style hinges made for inset doors.

There are several types of pocket-door hardware on the market. The most common and least expensive resembles paired drawer slides. I use them only for doors up to about 30 in. high. Larger doors sag when extended and rub on the cabinet bottoms. Accuride® and Blum both make versions of this slide.

For larger doors, I use Accuride model 1332 hardware. The 1332 uses a set of cables like those on a draftsman's parallel rule to keep the hinges perfectly aligned.

Accuride suggests a ⅛-in. margin between the doors and the cabinet, but I think 1⁄16 in. looks better. In entertainment centers with pocket doors, I hang a shelf unit from the top of the cabinet (see the top right photo above). This provides a pull-out shelf above the TV for the VCR while leaving space for the pocket doors to slide.

Making the Best of Extendable Turntables

The story at the beginning of this chapter doesn't always discourage my clients from turntables. If they insist, I buy the heaviest-duty turntable I can find, usually the KV® 1385. The cost difference between this 200-lb. rated turntable and a 150-lb. rated turntable is less than $5. I fasten turntables to the cabinet with ¼-in. #20 bolts and washers. Wood screws will eventually strip out, dropping that $900 TV to the floor.

A more stable platform for extending the TV from the cabinet is a lazy Susan affixed to a shelf. The shelf is mounted on 200-lb. rated full-extension drawer slides. This set-up is much steadier, but it can't be used on entertainment centers that have pocket doors. The drawer slides must be screwed to the sides of the cabinet and would leave no room for the pocket doors.

Brian Wormington owns Acorn Woodworks in Otis, Massachusetts.

Bookshelf Basics

■ BY BRUCE GREENLAW

I t's hard to define the quintessential book-shelf. The one above my writing desk, for example—a plastic-laminated particleboard shelf supported by three inexpensive metal wall brackets—was quick to build and per-fectly suits my needs. The fixed shelf puts my reference books within arm's reach of my chair, and foam-padded steel bookends that I got at Wal-Mart® keep them from falling off. Thos. Moser Cabinetmakers' fur-niture-grade cherry bookcases (Thos. Moser Cabinetmakers, PO Box 1237, Auburn, ME 04211; 800-862-1973) (see the photo on the facing page), on the other hand, are de-signed to be heirlooms.

Between these two extremes lie a wide va-riety of shelving options. Basically, though, bookshelves are either housed in bookcases or supported by wall brackets (unless they're propped on milk crates or cinder blocks), and they're either fixed or adjustable. To my mind, the best shelving systems comple-ment their surroundings and don't droop when they're loaded with books.

This chapter contains food for thought on designing bookshelves, plus an appraisal of shelving materials and hardware (see "Sources" on p. 279). Of course, this infor-mation can be applied to virtually any type of shelving.

Measure the Books, and Size the Shelves

Standard paperback books are about 4 in. wide (from the spine to the outside edge), but my binders, portfolios, and biggest reference books are about 11 in. wide. Un-less shelves will be used for storing old LP records (which are 12¼ in. wide) or giant art books, you'll rarely need to make book-shelves that are more than 11 in. deep. Some shops make 8-in.- to 10-in.-deep shelves and let wide books overhang.

If a series of shelves will be fixed in a bookcase or on nonadjustable wall brackets, measure the heights of the books that will be stored on them and space the shelves ac-cordingly. Books range in height from about 6¾ in. tall for standard paperbacks to well over 12 in. tall, but most are in the 8 in. to 12 in. range. Don't forget to add ¾ in. to 1 in. of clearance above the books to allow fingers to grip them.

Shelf Standards and Hole-Mounted Clips Make Bookcases Adjustable

Unless they're required for structural integrity, fixed shelves are probably unnecessary in bookcases. Adjustable shelves take much of the worry out of planning, offer more long-term flexibility, and often result in more shelves fitting into a given space. The joinery for a bookcase without fixed shelves is simplified because you're basically just building a big box. Also, the supporting hardware for adjustable shelves is relatively inexpensive and easy to install.

Most shops don't use run-of-the-mill metal shelf standards to support exposed shelving. But metal standards are stronger than most alternatives, and I think they look okay if their color complements the surrounding bookcase. The standards are screwed or nailed in pairs to bookcase sides. They can be mounted on the surface, but most also can be recessed into dadoes for a more refined look (see the bottom right photo on p. 276). Barbed plastic standards

Bookshelves can transform a room into a library. Open shelves provide plenty of easily accessible room for books and collectibles. Glass-panel and raised-panel doors provide more secure shelf space, and related supplies can be stored in lockable drawers.

Sizes, Standards and Brackets

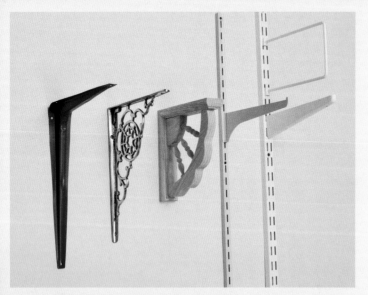

Bookshelves can be quickly and easily installed on wall brackets. Fixed brackets are screwed directly to the wall; adjustable brackets mount on slotted standards and can be moved up or down.

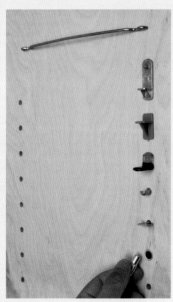

Shelf pins are inconspicuous. Shelf pins are less visible than shelf standards. Top to bottom: wire clip, which hides inside a groove at the end of a shelf; spring-loaded locking clip that prevents shelves from sliding or tipping; plastic clip; cushioned metal L-support; ornamental solid-brass pin; zinc-plated steel paddle; "library" pin with sleeve.

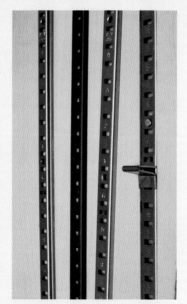

Metal standards are strong but conspicuous. Although metal shelf standards come with various disguising finishes, they tend to be more visible than hole-mounted shelf supports. Most standards can be recessed (at right) to make a finished appearance and to minimize gaps at shelf ends, or they can be surface-mounted.

that are simply pressed into dadoes (no nails or screws are required) also are available. But I've heard that these standards can be difficult to align, resulting in wobbly shelves.

Shelf pins are generally regarded as a step up from shelf standards (see the bottom left photo). These pins are strong enough for most applications, and they're less conspicuous than shelf standards. Shelf pins typically have ¼-in.-dia. shanks that fit into holes of the same diameter. The holes are spaced about 1 in. to 2½ in. o.c., and they're positioned so that shelves overhang the pins about 1 in. to 2 in. front and back.

Drilling shelf-pin holes can be tedious work, but a drilling template makes the job easier. A simple one can be made from a piece of tempered pegboard, which has ¼-in.-dia. holes spaced 1 in. o.c. (see the top photo on the facing page). Using a drill bit with a drill stop mounted on it speeds the work and prevents the bit from boring through the workpiece.

Pegboard jigs don't last long, though. Commercial jigs are a better choice for production work. The acrylic shelf jig sold by The Woodworkers' Store (see the bottom photo on the facing page) has oversize holes that guide a self-centering ¼-in. Vix bit (or a 5mm Vix bit for metric work). The jig is only 19 in. long, but it can be indexed with a shelf pin to bore any number of holes accurately. The jig costs $16, but the Vix bit adds another $35. Another shelf-drilling jig is made by Veritas® tools and is available through mail-order outlets (Veritas Tools, 12 E. River St., Ogdensburg, NY 13669; 800-667-2986).

Woodworking and hardware suppliers sell a variety of shelf pins and supports (see the bottom left photo). Trussed plastic shelf pins are the least expensive and the most obtrusive. Metal L-supports are stronger and more subtle than plastic pins. Padded L-supports cushion glass shelves and help protect fragile finishes. Locking pins made of plastic or metal help prevent shelves from sliding and tipping. They allow bookcases to be shipped

Templates simplify boring of shelf-pin holes. Drilling jigs make it easy to bore shelf-pin holes in parallel rows in bookcase sides. This disposable shopmade jig is tempered-hardboard pegboard, which has ¼-in.-dia. holes spaced 1 in. o.c.

Commercial hole-drilling jigs are durable. The acrylic jig shown here (sold by The Woodworkers' Store) has over-size holes that guide a self-centering Vix bit. A shelf pin is used as an indexing pin to extend the range of the jig.

with shelves installed, and they help anchor shelves in earthquake country.

To my eye, the best-looking shelf supports on the market are spoon-shaped pins made of nickel or brass. Several shops I know of bore oversize holes and tap special metal sleeves into the holes to support these pins. The sleeves help prevent the holes from deforming, and they lend an air of refinement to unused holes. Dave Sanders & Company, Woodworker's Supply, and others sell matching spoons and sleeves. Shelf pins can also be hand-carved, cut from dowels, or even fashioned out of brass brazing rods. For invisible shelf support, wire clips are the best choice. They're inserted into holes bored into the sides of bookcases, and they hide inside sawkerfs cut into the ends of shelves.

Wall-Mounted Shelf Brackets Can Be Fixed or Adjustable

Truth is, most of the bookshelves I've put up sit on plain metal wall brackets. The brackets install quickly with hollow-wall anchors or by screwing them directly to studs. Deluxe shelf brackets that install just as easily also are available. Two examples are the brass brackets sold by Renovator's Supply and the oak gingerbread brackets sold by The Woodworkers' Store (see the top photo on the facing page).

For adjustable wall support, most hardware stores sell single-slotted metal standards that screw to walls and carry flimsy

Designing Shelves for Looks and Function

Aprons stiffen the shelves. Putting a solid-wood apron on a shelf increases its load-carrying capacity. The sagging top shelf is ¾-in. particleboard. The virtually straight bottom shelf is ¾-in. particleboard with a 1x2 apron glued to the front edge.

metal brackets that hook into the slots. But heavier-duty, twin-slotted systems (which have two rows of slots and hooks instead of one) are also available. The system sold by The Woodworkers' Store includes special screws that prevent shelves from sliding off the brackets and bookends that clip to the standards.

Dave Sanders & Company has the best selection of wall-mounted standards and brackets I know of. One type is mounted to studs before drywalling so that only a slim slot is visible afterward.

Well-Designed Shelves Don't Sag

Most cabinet shops use simple rules of thumb to determine shelf spans. Generally (when ¾-in.-thick stock is used to support heavy reference books), particleboard and medium-density fiberboard (MDF) shelves span up to about 20 in.; softwood and plywood shelves span up to about 34 in.; and hardwood shelves span about 36 in. Shelves

that carry paperbacks will span significantly farther.

For more predictable control over shelf sag, some shops use the shelf-deflection table published in *Architectural Woodwork Quality Standards* (available for $50, or $5 for members, from the Architectural Woodwork Institute (AWI), 13924 Braddock Road, Suite 100, Centreville, VA 22020; 703-222-1100). The table lists the uniform loads that cause various unfixed 8-in.- and 12-in.-wide shelving materials to deflect ¼ in. when spanning 30 in., 36 in., 42 in., and 48 in.

AWI's chart points out a number of ways to beef up shelves to increase spans or to support unusually heavy loads. For instance, gluing a ⅛-in.-thick edgeband to a ¾-in.-thick particleboard shelf increases the shelf's load-bearing capacity by about 15 percent. Veneering the particleboard on two faces and one edge with ⁵⁄₁₀₀-in.-thick plastic laminate increases its capacity by about 200 percent. Gluing a 1x2 solid-wood apron to the front edge boosts it by a whopping 300 percent to 400 percent (see the sidebar at left). Greg Heuer, AWI's director of member services, notes that applying an apron to the back of a shelf, even pointing up (so that it hides behind books), gives the same results as a front apron. Although most people would consider a shelf deflection of ¼ in. to be excessive, halving the values listed in the chart would produce a less noticeable deflection of ⅛ in.

Several other strategies can be used for increasing shelf spans or load-bearing capacities. The most obvious is to use thicker shelves. Shelves made of 2x lumber, or two layers of ¾-in. plywood, for example, will span at least 48 in. Applying a half-round molding to the front edge of layered plywood creates a solid bullnose shelf.

For a significant span, consider building a torsion-box shelf, which works somewhat like a box beam or hollow-core door. These shelves consist of a grid of wood or plywood strips glued between two plywood skins. The biggest torsion-box shelf that I've heard of is 4 in. thick and 27 ft. long. Fixed shelves

Sources for Bookshelf Hardware

Here's a short list of some useful companies to know about if you're in the market for bookshelf hardware.

Architectural Woodwork Institute
13924 Braddock Road, Suite 100
Centreville, VA 22020
703-222-1100
www.awinet.org

Dave Sanders & Company Inc.
19-40 45th St.
Astoria, NY 11105
718-726-5151
www.davesanders.com

Sells an impressive assortment of shelf standards, brackets, and shelf pins, including Magic Wire concealed shelf supports and shelf pins with matching sleeves that reinforce pinholes.

Knape & Vogt Manufacturing Company
2700 Oak Industrial Drive NE
Grand Rapids, MI 49505
800-253-1561
www.knapeandvogt.com

Makes cabinet and shelving hardware, including twin-slotted, wall-mounted standards and brackets, shelf pins, and heavy-duty metal wall brackets that support up to 1,000 lb. per pair.

Lee Valley Tools Ltd.
PO Box 1780
Ogdensburg, NY 13669
800-871-8158
www.leevalley.com

Rakks/Rangine Corporation
330 Reservoir Street
Needham, MA 02494
800-826-6006
www.rakks.com

Makes storage systems, including the Rakks Shelving System: extruded-aluminum, wall-mounted shelf standards with locking, infinitely adjustable aluminum brackets.

Renovator's Supply
PO Box 2515, Dept. 9898
Conway, NH 03818
800-659-0203
www.rensup.com

Sells ornamental brass shelf brackets.

Woodcraft Supply
PO Box 1686
Parkersburg, WV 26102
800-225-1153
www.woodcraft.com

Sells assorted shelf pins, including solid-brass pins and brass sleeves that reinforce pinholes.

Rockler Woodworking and Hardware
4365 Willow Drive
Medina, MN 53470
800-279-4441
www.rockler.com

Sells a range of knockdown fasteners, standards, brackets, and pins; the best assortment of edgeband I know of; and tools and accessories.

Thomas Moser Cabinetmakers
PO Box 1237
Auburn, ME 04211
800-862-1973
www.thomasmoser.com

---

in bookcases can be reinforced by putting a back on the case and fastening the back to the shelves, or by fastening intermediate support posts to the front edges. Adjustable bookcase shelves can be reinforced at the back with shelf standards and brackets or with hole-mounted shelf pins.

If you doubt the ability of a proposed shelf to support a given load, prop a sample shelf on a pair of blocks, load it with the weight it will carry and check the sag.

Recessing and Trim Make Shelving Look Built In

Louis Mackall, owner of Breakfast Woodworks Inc. in Guilford, Conn., tells me that the best way to integrate shelving into a room is to recess it at least partway into a wall (see the photo on p. 280), even if this procedure requires furring out the wall. According to Mackall, recessed shelves not only look better than projecting shelves, but they also appear to add space. Projecting shelves appear to subtract space. Continuing a room's base and crown moldings (if there

Bracket-mounted bookshelves can look built in. Philadelphia woodworker Jack Larimore applies fancy aprons to basic wall-mounted bookshelves to produce ornate library storage that is available for a modest price.

Recessed shelving doesn't intrude. Breakfast Woodworks in Guilford, Conn., recesses shelving to blend it with the architecture and to create the illusion of space.

are any) around a bookcase also helps to unite the bookcase visually with the room.

San Francisco woodworker Scott Wynn considers horizontal details in a room when sizing built-ins. If window head casings are a prominent feature, for instance, he'll make the built-ins the same height as the head casings. Wynn also makes bookcases supported by deeper base cabinets. The cabinet tops serve as oversize shelves that support art books. Wynn cautions that bookcase partitions must be placed directly over cabinet partitions or the shelves will sag.

Except for furniture-grade pieces such as Thos. Moser's, most bookcases are boxes with routed edges or applied trim. The boxes can have butt joints held together with nails or screws in concealed locations, or with biscuits or plugged screws in exposed loca-

tions. Special knockdown fasteners can also be used, allowing units to be dismantled and reassembled. Tall bookcases should have ¼-in. hardboard or plywood rabbeted into and tacked to the back for stability, a 1x nailer at the top for attachment to walls, or both.

Bookshelves don't have to nest in bookcases to look good. Philadelphia woodworker Jack Larimore has designed and built economical wall-mounted shelving that looks like pricey built-in furniture. The shelves are supported by metal shelf standards, but they're dressed up with decorative aprons. Ornamental metal and wood brackets can also be used to enhance the appearance of wall-mounted shelving (see the left photo on p. 276).

Solid Wood Is Beautiful but Unstable

Thos. Moser builds most of its bookcases out of ¾-in.-thick American black cherry. At the other extreme, I've used low-cost #3 knotty pine for a number of shelving jobs because it's relatively strong and inexpensive. Unlike some shelving materials, solid wood doesn't require edgeband, and edges can be detailed easily with a router or a handplane.

Unfortunately, solid wood is unstable. It shrinks as it dries, and it expands and contracts across the grain in response to changes in humidity. It also tends to warp and cup more than sheet stock, causing adjustable shelves to seesaw.

One nifty idea I've seen is to use bull-nose hardwood stair-tread stock for shelves. It usually comes in red or white oak, it's often glued up out of narrow strips for dimensional stability, and it's a sturdy 1 in. to 1¹⁄₁₆ in. thick. Millworks and many lumberyards sell it, and when you factor in presanded finish and ready-made edge treatment, the price is reasonable.

Hardwood Plywood Is an Industry Favorite

Most shops I spoke with shun solid wood in favor of sheet stock such as hardwood plywood, particleboard, or MDF (see the photo below right). Sheet stock is more stable than solid wood, and it normally can be turned into shelving parts faster because it doesn't have to be flattened and straightened as solid wood often does.

Hardwood plywood is available with a veneer core, an MDF core, a particleboard core, or a lumber core. Veneer-core plywood is strong, has excellent dimensional stability, holds screws well, and is widely available (especially in birch, red oak, and lauan). But it can have voids in the core that could occasionally cause a hole-mounted shelf pin to sag or to fall out.

MDF-core and particleboard-core plywood don't have voids, but they also don't hold screws as well as veneer-core plywoods do, although using particleboard screws (which have a deep, coarse thread) helps. These panels are also heavy, weighing almost 100 lb. per ¾-in.-thick sheet. This

Choosing the Right Materials

Cover exposed plywood edges with edging. Options include (top) T-molding; pine screen bead; plastic T-molding; peel-and-stick hardwood edgeband (outer roll); and iron-on polyester edgeband (inner roll).

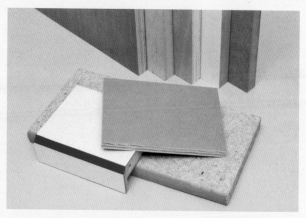

Sheet stock is a stable and uniform choice for shelving. Options range from hardwood-veneer plywood to medium-density fiberboard panels with a durable melamine surface.

weight not only makes them awkward to machine, but it also makes them a questionable choice for portable bookcases.

Nevertheless, some shops use MDF-core plywood exclusively for their bookshelves. They like its solid core and its cost, which is about 5 percent to 20 percent less than veneer-core plywood. Georgia-Pacific's Fiber-Ply® core plywood and Columbia Forest Products' Classic Core® plywood are compelling hybrids. They look like conventional veneer-core plywood, but they have homogenous layers of fiberboard instead of solid-wood plies directly beneath the face veneers.

Lumber-core plywood is strong, it holds screws tenaciously, and it is supposed to be voidless. But today's lumber-core plywood, much of which is imported, can have shrinkage voids in the core. Good-quality lumber-core plywood is expensive and can be hard to find.

Regardless of the type of plywood used, exposed edges need to be covered with moldings or edgeband (see the left photo on p. 281). Moldings can be anything from pine screen bead to solid-wood or plastic T-moldings that fit into grooves in panel edges. Edgeband is peel-and-stick, iron-on or glue-on veneer that comes on a roll. Solid-wood, polyester and even metallic-foil edgebands are available.

Particleboard and MDF Work Well for Paint-Grade Work

Particleboard and MDF are hard to beat for paint-grade shelving systems. They're relatively inexpensive and stable, and they don't require edgeband. Most lumberyards sell ¾-in. industrial-particleboard shelving with filled bullnose edges. Available in many widths—typically 8 in. or 12 in.—precut shelving is quick and convenient.

Called the "Buick of particleboard" by Albuquerque woodworker Sven Hanson, MDF is made of highly compressed wood fibers instead of particles, resulting in panel surfaces that are as flat as glass and edges that machine beautifully without chip-out (machining produces clouds of fine dust, though, so wearing a high-quality dust mask is a must). The Home Depot in my area sells ¾-in. MDF for about $20 per sheet, which is a bargain.

One problem with MDF and particleboard panels is that unsealed edges drink paint, resulting in an uneven finish. You can seal MDF edges with two coats of water-based polyurethane finish, but quick-drying sanding sealers and white PVA glues diluted 20 percent with water will also work.

Paint-grade plywood, called MDO (medium-density overlay), can also be used for painted shelving. This exterior-grade veneer-core plywood is coated with a resin-treated paper that makes a superb, smooth substrate for paint. MDO can also be taped like drywall, so it makes a good material for built-in shelving.

MDF and particleboard can also come veneered with melamine or plastic laminate. Melamine is a thermally fused, resin-impregnated sheeting that resists abrasion, stains, heat, and chemicals, and it's available in a variety of sizes, patterns, and colors. The Home Depot sells melamine panels that are ripped to standard bookcase widths and that are predrilled with ¼-in. holes that accept standard shelf pins. Melamine can be tricky to work with because it chips easily. But it's a versatile material, and it has a hard prefinished surface.

Bruce Greenlaw is a contributing editor of Fine Homebuilding. *Photos by the author.*

Installing Kitchen Cabinets

■ BY KEVIN LUDDY

In Jonathan Swift's novel *Gulliver's Travels*, the Lilliputians were at war with their neighbors over whether a boiled egg should be cracked from the pointy end or from the round end. At teatime, the British squabble over whether the tea or the milk goes into the cup first. Likewise, on job sites I've seen carpenters argue vehemently about which kitchen cabinets should be installed first, lowers or uppers. There are good reasons for each choice, but you'll have to read on to find out which way I prefer.

Check Everything before Installation Begins

With this kitchen (and with all of my kitchen jobs), I received an information packet from the designer. This packet included the floor plan and elevations that showed backsplash and crown-molding heights, as well as countertop dimensions and the specs of the four built-in appliances. All this information helps me to bid the job accurately, but I also use it during installation. My first task at the site was to check the stock numbers of the cabinets against the floor plan to make sure the order was complete and correct. I also looked at the cabinets for any obvious damage and arranged them in order of installation and by area (e.g., the island cabinets all together).

The Starting Line

The floor is first checked for level (1). The cabinet height is measured from the high point, and a level line is drawn. The same height is measured from the low point (2), and a second line is drawn (3) to find the difference and to check the first line. The first base cabinet is set and shimmed to the top line (4).

Next I measured everything in the kitchen, checking my measurements against the floor plans and elevations. I checked wall heights and lengths, locations of lights and outlets, and window and door locations; I also checked the walls for plumb and straight. The only major problem I found in my investigation was an out-of-place outlet, and lucky for me, an electrician was on site that day.

Two Lines Set the Cabinet Height

After checking the floor for level (see photo 1). I measured up the height of the base cabinets (in this case 34½ in.) from the high spot for each run of cabinets and drew a level line across the wall with a 4-ft. level. A laser level would also work well here. I don't recommend snapping chalklines,

Joining Cabinets

Face frames are clamped flush (5) and then screwed together. A framing square clamped to the cabinet (6) keeps tall cabinets square to shorter ones. Tall cabinets are plumbed (7) and then screwed to the wall. Cabinets are kept level while they're being joined (8).

though. They're usually too fuzzy and not always level.

Next, I found the low spot for each run, measured up the same distance (see photo 2) and drew a second line (see photo 3), which gave me the range of floor error and checked the accuracy of my first line. If the lines are parallel, I'm all set. If not, I go back and try it again. The finished kitchen floor (½-in. tile over ⅜-in. subfloor) had not been installed, so I could count on it to hide shims. (Otherwise, I would have had to undercut, or scribe, every base cabinet, which is time-consuming.) The top line then became my guide for cabinet level.

I then marked the stud locations along the level line. I also located critical points such as the window centerline. Some carpenters mark the location of each cabinet along the level line, but I usually scrunch the exact locations by fractions as I go to make everything fit, which would render those marks wrong.

The Base Cabinets Go in First

Installing base cabinets first takes a little more care, but I use the lowers to help with installing the uppers. Besides, most bad surprises show up when installing the lowers.

For this kitchen, I began with the lazy-Susan corner cabinet. I transferred the stud locations to the back of the cabinet and drilled screw holes through the hanging strip for every stud location.

Then I placed the cabinet in the corner, shimmed it plumb and level (see photo 4 on p. 284), and drove a #10 by 2½-in. pan-head screw through each hole. I steer clear of using drywall screws when I'm attaching cabinets because they have little shear resistance. Then I laid out and drilled the screw holes for the cabinet to the left of the corner and slid it into place.

To join the two cabinets, I first clamped the face frames together, lining up the faces and the tops flush. Next I shimmed the cabinet plumb and level, and joined the face frames with screws through predrilled holes. I finished by screwing the cabinet to the wall.

This kitchen had a full-height wall-oven cabinet next in line. To attach the cabinet, I joined the face frames as before (see photo 5 on p. 285) but clamped a framing square to the top of the base cabinet to keep the oven cabinet square to its neighbor (see photo 6 on p. 285). I then plumbed the other side of the oven cabinet (see photo 7 on p. 285) and drove screws top and bottom to hold it in place.

Fridge Cabinet Is Assembled on Site

On the other side of the corner, I leveled and installed a double-drawer base cabinet (see photo 8 on p. 285). The fridge cabinet that came next had to be built on site from factory-supplied parts.

I began by setting the overhead cabinet on its side. Next I positioned one side panel on the overhead cabinet, supporting its loose end on a base cabinet. After lining up the top and front edges, I screwed through the panel and into the cabinet.

After carefully flipping over the unit, I installed the opposite side panel (see photo 9 on the facing page), this time using trim screws. With the unit assembled, I lifted it and rotated it into place (see photo 10 on the facing page). Like most refrigerator cabinets, this one was deeper than the base cabinets. I plumbed and leveled the cabinet and then screwed the left side into the face frame of the adjoining base cabinet.

Holding the unit plumb, I screwed the overhead cabinet into the studs. To stabilize the sides and to keep them at the proper width, I screwed two spreader cleats to the wall, one at the floor and another at the height of the base cabinets (see photo 11 on the facing page). A trim screw anchored the side panel to the cleats. The finish floor will lock the bottoms of the panels firmly in place.

Bevel Cut Makes a Tight Scribe

This kitchen had two other cabinet runs, a freestanding island, and cabinets along an outside wall for the kitchen sink, dishwasher, and trash compactor. The kitchen-sink base had to fit over a toe-kick heater and accept an unusual sink, so I assembled it from parts supplied by the cabinet company (see the sidebar on p. 288).

The sink base had a fixed width, so I marked out its dimensions centered on the window. Next I double-checked all the clearance specs and left the correct space on both sides of the sink base for the appliances. On the right side of the dishwasher, I installed a wine-rack end panel screwed to a 2x cleat on the wall.

The cabinet on the other side of the trash compactor had a preattached filler piece that had to be scribed to make the cabinet fit properly. To mark the scribe, I first set the cabinet in place and shimmed it level and plumb. Then I set my compass scribes and marked the cut on the filler piece (see photo 12 on p. 289).

Building the Fridge Cabinet

The refrigerator cabinet is made from two side panels joined to a top cabinet (9). The cabinet is carefully positioned (10) and attached to its neighbor. Wall cleats hold the sides at the proper width (11).

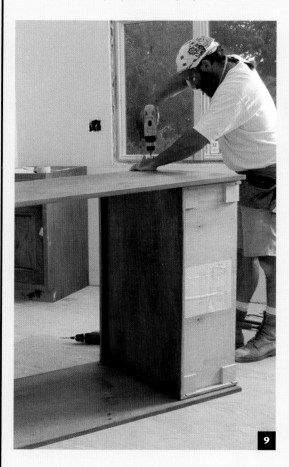

I try to put a slight bevel on all my scribe cuts, so I tipped the saw table slightly as I followed the line (see photo 13 on p. 289). Masking tape on the bottom of the saw helps to protect the cabinet. The bevel cut lets the cabinet fit more tightly against the wall (see photo 14 on p. 289). If the cut needs fine-tuning, I do it with an electric grinder. When I was happy with the fit, I screwed the cabinet into place. Note that with a wider filler piece, a wall cleat may be needed for attachment.

Blocks Anchor the Island in Place

Islands and peninsulas are special situations because the cabinets don't attach to walls. For this island, I began by snapping chalklines on the floor, laying out the full perimeter of the island.

One side of this island had a standard toekick, but the other side was to be covered with a solid beadboard panel. On the toekick

A Sink Cabinet from Factory-Supplied Parts

While writing a song, a friend once asked me how many Ps there were in "obstacle," as though it rhymed with popsicle. Of course, there are no Ps in "obstacle," but there were many obstacles to using a stock cabinet at the sink location in this kitchen project.

In addition to the plumbing pipes, it was to receive a specialty sink, and it sat over a toekick heater vent. With all these variables, the designers opted to build the cabinet on site from factory-supplied parts.

I started by building the toekick of 2x stock ripped to a height that would fit the vent cover supplied to me. It was slightly taller than the standard toekick, but the difference would be seen only by the bristles on the kitchen broom. With the appropriate gap left for the vent, I squared and leveled the base, centered it under the window, and glued and screwed it to the floor (see the photo below).

The base platform went on next, followed by the sides (see the top right photo). I cut the lower front corners of the side pieces to match the toekick on the rest of the kitchen. A 1x wall cleat anchored the sides in back.

Building the sink base from parts meant not having to drill holes for the plumbing, but when drilling is necessary, I measure the vertical location from my level line on the wall. I take the horizontal location from the adjacent cabinet and transfer this information to the back of the sink base, remembering to subtract for any face-frame overhangs. Next, I bore halfway through from the backside and then finish the hole from the inside.

The final step was installing the finished front, which was glued and tacked to the sides (see the bottom photo below). It looked a little funny when I was done, but I knew that with the countertop, sink, and appliances in place, you'd never be able to tell that the cabinet was built on site.

Scribing an Attached Filler

Cabinets with attached fillers have to be scribed to fit. With the cabinet in place, the cut is traced onto the filler (12). Next, the cut is made with the saw tipped to create a bevel (13), which lets the cabinet fit tight against the wall (14).

side, I subtracted the toekick depth and the thickness of the cabinet wall and snapped a second line. I also marked where the cabinets were joined together. I then glued and screwed 2x blocks to the floor at the ends of the island and along the toekick, leaving plenty of space where the cabinets would be joined (see photo 15 on p. 290).

Next I ran a bead of construction adhesive on the blocks where they would contact the cabinets and then slipped the cabinets over the blocks. The cabinets were then clamped and screwed together in the front and back (see photo 16 on p. 290), keeping the whole assembly level and square. I screwed the cabinets to the blocks along the toekick (see photo 17 on p. 290) where the screws would be hidden and drove the trim screws through the end panels to hold them in place until the glue cured. The beadboard panel went on next (see photo 18 on p. 290), and the island was solid as a rock.

Upper Cabinets Ride Piggyback on the Bases

With all the lower cabinets in place, the final step before moving to the upper cabinets was installing countertop support cleats as needed. The first upper cabinet I set in was the corner unit that was to rest on the countertop. I blocked the corner unit to the thickness of the countertop and made sure that it was absolutely plumb and level. I also checked the distance to the other cabinets. The corner cabinet was left loose with the idea that I could slide it up and out of the way when the countertop was installed and then drop it down for a precise fit on the granite counter.

For this kitchen, the refrigerator cabinet and the oven cabinet set the height of the upper cabinets. I marked the stud locations

Assembling the Island

After chalklines are snapped for the island, 2x blocks are glued and screwed inside the perimeter (15). The cabinets are then glued to the blocks and joined front and back (16). Screws driven into the blocks from the front (17) will be hidden by the toe kick. A finished panel is then glued and screwed to the cabinet backs (18).

on the hanging strip of the first upper cabinet and drilled holes through to the inside of the cabinet (see photo 19 on the facing page).

Next I cut two 2x riser blocks to the exact distance between the upper cabinets and lower cabinets. I screwed the blocks to the wall where the screw holes would be hidden by the backsplash (see photo 20 on the facing page), which let the base cabinets take the weight and set the level. I then placed the predrilled cabinet on the blocks and pushed it into place against the wall.

I drove screws through a couple of the holes to hold the cabinet in place temporarily while I checked for plumb and level. Shims were added where needed, and then I

Upper Cabinets Get a Boost

Stud locations are marked on the back of the cabinet, and holes are drilled (19). Blocks cut to the distance between the uppers and lowers and screwed to the wall (20) hold and level the cabinet until screws are driven into the predrilled holes (21). After plumbing and shimming are done, extra screws are driven through the hanging strip and through the face frame.

drove in permanent screws through all the predrilled holes (see photo 21). I also predrilled and drove in screws along the bottom inside of the cabinet and through the face frames.

I rechecked the corner cabinet for plumb and followed the same installation procedure for the rest of the uppers. At the window, I set the wall cabinet to the right of the window and then scribed the left-hand cabinet so that it fit at the same distance from the window.

Finishing Up

As the upper cabinets go in, I pay extra attention to the doors to make sure that plumbing and leveling haven't caused them to rack. If I do notice a racked door, I try to cheat the cabinet a little to compensate, or I fine-tune the door after all the cabinets are

in. I finish off the kitchen by applying the toekick, crown molding, and knobs.

These kitchen cabinets came with holes for the knobs already drilled. But if this isn't the case, I double-check with the clients for the exact knob locations before mounting the knobs or pulls. I putty all the nail and trim-screw holes, but usually save any touchup until after the countertops are installed. All doors are given a last check for swing and fit, and I check shelves, drawers, lazy Susans, etc.

I give the room a quick sweep, which keeps me in the good graces of the clients and contractors and ensures that I've rounded up all my tools. Now we're finished. Let's eat!

Kevin Luddy runs Keltic Woodworking, a custom-carpentry and cabinet-work business in Wellfleet, Massachusetts.

Simple Frameless Cabinets Built on Site

■ BY JOSEPH B. LANZA

I have to admit I was a bit slow to catch on to the advantages of frameless cabinets. I had problems with the edgebanding and the European-style cup hinges that inevitably go along with frameless cabinets. When I first tried the hinges, I found them awkward to use. Setup and layout meant deciphering arcane diagrams with odd, unfamiliar dimensions. The much-touted adjustability of the hinges seemed to require an awful lot of adjusting and readjusting. Frustrated, I concluded that this cabinet-hinge combination was best suited for factory production or, at the very least, a shop dedicated to the 32mm cabinetmaking system that originated in Europe.

But that was quite a few years ago. Now, after a series of improvements to both the hardware and the user, I find myself using these hinges all the time. The clip-style hinge, now made by several manufacturers, is a big improvement over earlier versions. This hinge makes it possible to hang and remove doors quickly, without tedious readjustment every time. I incorporated these

hinges into a hybrid system for building face-frame cabinets with inset doors.

These hinges were so much better, they encouraged me to give frameless cabinets another try. They have a clean look that seems especially well suited to a spare architectural style. I looked for a way to build them on site quickly and easily, and the answer was simple: Make cabinet boxes just as I'd been making drawer boxes for years.

Using Baltic-Birch Plywood Eliminates Edgebanding

For the little guy, edgebanding has always been the biggest headache in building frameless cabinets. The commonly available iron-on edgebanding just doesn't cut it for me, and ripping, applying, trimming, and sanding thin strips of solid stock is a huge pain in the neck. Because I make most of my drawers from ½-in. Baltic-birch plywood with tongue-and-groove corner joints, I realized I could make cabinet boxes the same

way with ¾-in. stock. Unlike ordinary hardwood plywood, Baltic birch is all birch with no softwood and no voids in the core. It is made up of thin veneers of alternating grain, which makes for a pleasing striped pattern when the edges are sanded or routed. Edgebanding can be avoided entirely.

As with cup hinges, Baltic birch can take a bit of getting used to. It comes in 5-ft. by 5-ft. sheets, so if you don't have long arms, a plastic panel lift will come in handy. Baltic birch tends to warp if left unused and unbraced, so if it looks as if it has been around the lumberyard for a while, you may want to pick from the middle of the stack. It also is

notoriously out of square, so you will want to check for (or make) a straight edge before you rip it. The sheets also tend to vary a bit in thickness. I've never had a problem with sheets from the same lift, but I sometimes run into a mismatch when mixing leftover pieces with a new delivery.

That said, it is great, and I love to work with it. Here in Massachusetts, Baltic birch is stocked in quite a few lumberyards and at nearly all plywood suppliers. The price for a sheet of ½-in. stock is around $25 and about $35 for ¾ in., but you can usually get a substantial price break if you can buy a full lift.

A tablesaw cutoff box ensures that cabinet parts will be square. When working with Baltic birch, don't assume that factory edges are square.

Strengthening a Tongue-and-Groove Joint

Orienting a cabinet bottom or shelf incorrectly (top) makes a weak joint. The shelf is susceptible to splitting and will behave as if it were the thickness of the tongue. Putting the joint together with the thick part of the shelf facing up (bottom) is the stronger option.

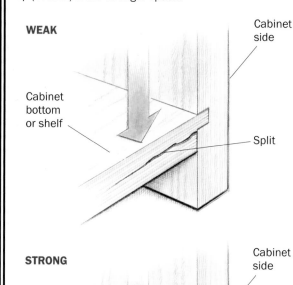

WEAK

Cabinet side

Cabinet bottom or shelf

Split

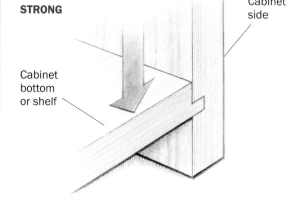

STRONG

Cabinet side

Cabinet bottom or shelf

Lay Out Cabinets on a Story Pole, Then Cut All the Parts

My first step in building cabinets always is to make a story pole, which is a piece of stock about 2 in. wide and as long as the longest dimension in the kitchen. I lay out everything on the pole. This procedure gives me a chance to make my mistakes at full scale and to check my measurements on site before building the cabinets. I mark all cabinet sides, tops and bottoms, and joints directly on the stick. I make a cutlist from the stick, then rip and crosscut all the cabinet parts (see the photo above) on a tablesaw.

Because there is no face frame to add strength and stiffness to the carcases, the corner joints have to be solid. There are a number of methods that work well—biscuits, screws, knockdown fasteners—but I like the tongue-and-groove joint. It is easy to set up on a tablesaw with a dado head, quick to cut and easy to assemble, and it gives lots of gluing surface. I make the cabinet sides full length and then cut the grooves in them, being sure to orient the groove so that the joint will have maximum strength (see the drawing at left). The cabinet tops and bottoms get the tongues.

The groove is ¼ in. wide and a strong ⅜ in. deep. I set up and cut all the grooves for the base cabinets at the same time. Then I move the fence 2 in. away from the dado head and run the wall cabinet sides, which have a 2-in. gap at top and bottom (top left photo, facing page). These gaps lower the top shelf, allowing the cabinet sides to rise above it as a curb. This gap also makes room for the bottom rails that conceal the recessed lights.

The dado setup for the tongues is the same for all cabinets, so after I get a good fit on a test piece, I cut all of them (see the top right photo on the facing page). To be strong, the joint must fit together snugly. Base cabinets don't have solid tops, just

Grooving cabinet sides 2 in. from their bottoms creates space to hide over-the-counter lights.

When cutting the tongues, a featherboard attached to an auxiliary fence makes the process safer and the tongues more consistent in thickness.

spreaders. To make them, I cut the tongue on a piece of scrap 6 in. to 7 in. wide (it's not critical), then rip that in half. These pieces go at the front and back of the cabinet to hold the sides together and to provide some support for the countertop. The cabinet bottom is, of course, solid. If the cabinets have backs, I cut them next (the cabinets in the photographs don't have backs). I make the cabinet rails from leftover ¾-in. stock.

Assembly Is Speedy with Screws or Air-Driven Nails

I assemble the cabinets with glue and, depending on what will be exposed, either screws or air-driven nails (bottom right photo). Because these base cabinets later received end panels, I used 16-ga. nails to assemble them and didn't worry about the holes. For the exposed sides of the wall cabinets, I used screws with integral washers. I used the same fasteners for the end panels and the exposed back of the kitchen island (see the photo on p. 293). I also used these fasteners to attach aluminum trim throughout the rest of the house.

Nails secure cabinet boxes while the glue dries. Check for square by comparing diagonal measurements before the glue sets.

Cabinets without backs are checked for square after they are assembled, then laid on the floor until the glue sets. Before installing any of these cabinets, I run a router with a ⅛-in. roundover bit around the inside and outside edges of the fronts. This little profile also fools the eye if the cabinets don't quite line up when they are installed.

After marking stud locations, the author hangs upper cabinets. Without backs and doors, these boxes are light enough for one person to hold in place as they are screwed to wall framing.

I hung the wall cabinets first (see the photo at left) while they were easy to reach, then set the bases. Building cabinets in a lot of old, out-of-level, out-of-square houses has gotten me in the habit of building bases separate from the cabinets. I level the bases first (bottom right photo), then set the cabinets on them. I like to spray the exposed toekicks with black paint before the bases are installed. Then I set the cabinets (see the bottom left photo) and attach finished ends and backs.

I made drawers like cabinet boxes and installed them before cutting parts for the doors and drawer fronts. Both the drawer fronts and doors are slabs of ¾-in. Baltic birch.

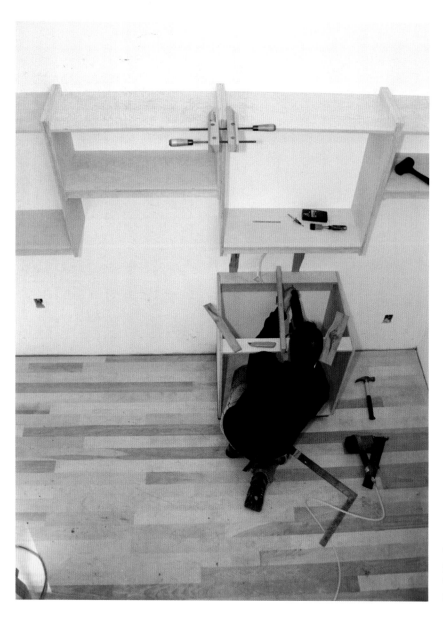

To make leveling them easier, the author installs the bases separately from the cabinets. With the bases secured, he then sets the cabinet boxes in place and screws them to the wall.

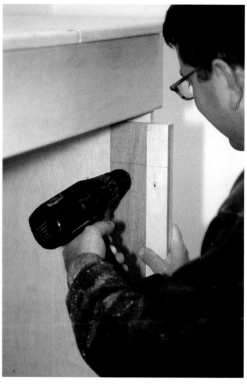

After testing the layout on scrap, the author drills 32mm holes in the backs of cabinet doors. Pencil marks on the drill-press fence indicate the door-edge alignment.

A jig locates screw holes for hinge plates inside the lower cabinets.

When it was time to drill big cup-hinge holes on the back sides of the doors, I set up a fence on my portable drill press. Pencil marks at 3½ in. on both sides of the bit space out the hinges equally (see the top left photo). I can drill one hole, slide the door across the fence until the edge reaches the other mark and drill the second hinge-cup hole. Before boring out a stack of doors, it is always a good idea to mock up a hinge on scrap stock to make sure that the layout is correct.

Before hanging doors, I ease the backs with a ⅛-in. roundover bit and the fronts with the ⅜-in. roundover to expose the edges of the birch plies. To hang doors, I made a plywood jig to drill holes for the hinge base plates (top right photo). When laid out correctly, the hinges snap easily together.

Joseph B. Lanza is a designer and builder who lives in Duxbury, Massachusetts.

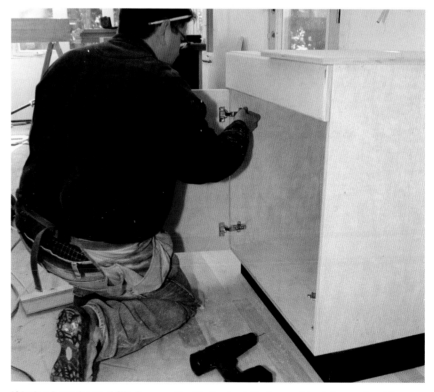

If holes have been drilled correctly, hinges installed on cabinet doors should snap easily into place on the base plates. These hinges allow the author to fine-tune the fit of the door.

Enhance Any Room with a Window Seat

■ BY GARY STRIEGLER

Like a hammock strung between two trees or a porch swing hanging from chains, a window seat is appealing well beyond its usefulness. More than just a place to sit, it represents the promise of leisure. If you have a window seat, you also might have the time to sit there and read.

I include a window seat in most of the homes that I build. It adds a cozy focal point to a room, plus useful storage beneath the hinged lid. Of course, you can forgo the operable lid for cabinet-style doors or even

drawers built into the face frame to gain access to the inside of the seat. That's a much more involved process. I prefer the traditional lid because it's faster and easier, and because it satisfies people's expectations.

The simple design shown here took me an afternoon to build and used less than $200 worth of materials.

Gary Striegler is a custom-home builder in Fayetteville, Arkansas.

Built-In Window Seat with Hidden Storage

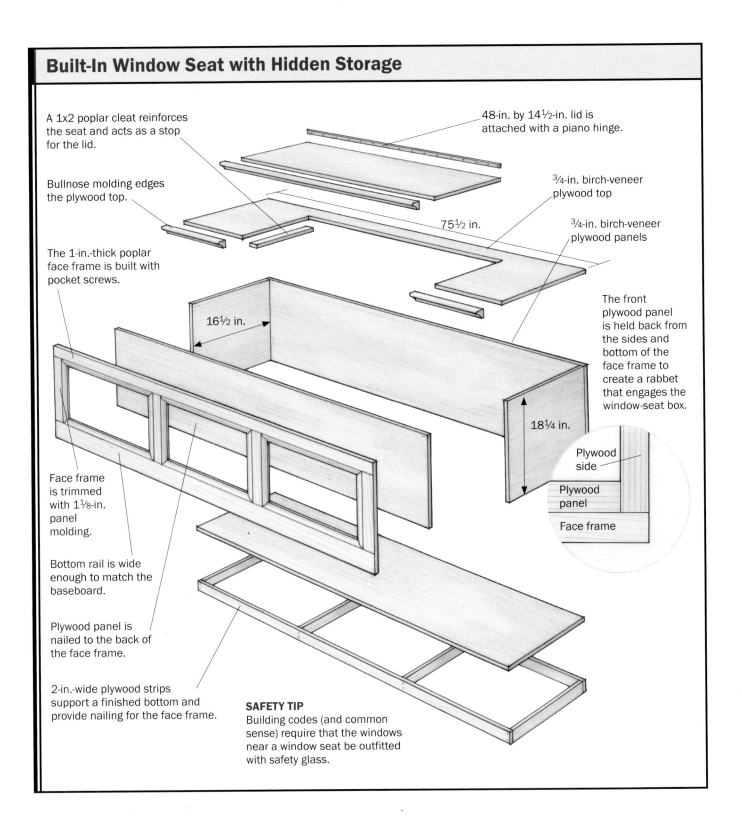

A 1x2 poplar cleat reinforces the seat and acts as a stop for the lid.

48-in. by 14½-in. lid is attached with a piano hinge.

Bullnose molding edges the plywood top.

¾-in. birch-veneer plywood top

¾-in. birch-veneer plywood panels

75½ in.

The 1-in.-thick poplar face frame is built with pocket screws.

16½ in.

The front plywood panel is held back from the sides and bottom of the face frame to create a rabbet that engages the window-seat box.

18¼ in.

Plywood side

Plywood panel

Face frame

Face frame is trimmed with 1⅛-in. panel molding.

Bottom rail is wide enough to match the baseboard.

Plywood panel is nailed to the back of the face frame.

2-in.-wide plywood strips support a finished bottom and provide nailing for the face frame.

SAFETY TIP
Building codes (and common sense) require that the windows near a window seat be outfitted with safety glass.

Build the Box in Place

Because the window seat is built in place and attaches to the walls, the construction is fast without any special joinery. The plywood panels are far more resilient than drywall, and they give the interior a finished look while acting as support for the top. A plywood floor hides any gaps under the side panels.

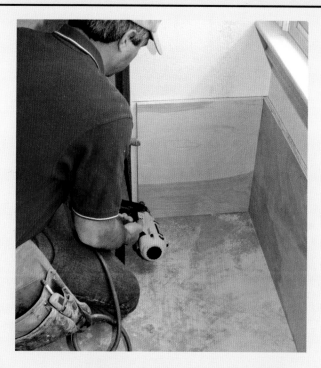

Secure the plywood panels to the studs with 15-ga. 2½-in. finishing nails, or 8ds if you're hand-nailing. The side panels are held back ¾ in. so that the face frame will be flush with the wall.

The floor doesn't need to be level. A frame for the window seat's floor is constructed of 2-in.-wide plywood strips and is nailed to the side and back panels.

Getting the Lid Just Right

If the alcove is square, making the top and its hinged lid from one piece of plywood is the most efficient approach but requires a plunge cut along the hinged edge, which is tricky to do. A plunge cut can be made with a table-saw or with a circular saw, which takes a little more time and skill. If the alcove is out of square, it will be tough to fit a one-piece top without scarring the drywall. In that case, a four-piece top (bottom drawing) is the best bet.

FIRST, FIND THE HINGE POINT
Ideally, you'll want a large lid that tilts back and stays open but doesn't rest on the window glass. Dry-fit the top and place a framing square so that the tongue (short leg) rests against the window's stool. Measure forward about 1 in.; this is where you'll hinge the lid.

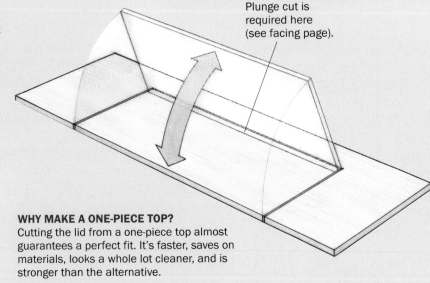

Plunge cut is required here (see facing page).

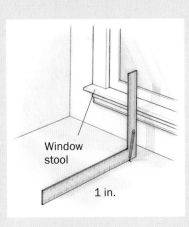

Window stool

1 in.

WHY MAKE A ONE-PIECE TOP?
Cutting the lid from a one-piece top almost guarantees a perfect fit. It's faster, saves on materials, looks a whole lot cleaner, and is stronger than the alternative.

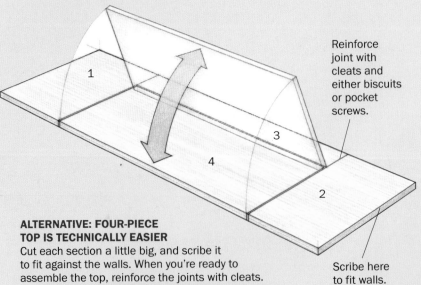

Reinforce joint with cleats and either biscuits or pocket screws.

Scribe here to fit walls.

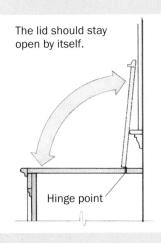

The lid should stay open by itself.

Hinge point

ALTERNATIVE: FOUR-PIECE TOP IS TECHNICALLY EASIER
Cut each section a little big, and scribe it to fit against the walls. When you're ready to assemble the top, reinforce the joints with cleats. It's important that the opening is square so that the lid opens and closes cleanly.

TWO WAYS TO PLUNGE CUT

Tablesaw Method

Don't lower the wood into the spinning blade; instead, raise the blade into the wood. With the saw turned off, raise the blade about 1 in., and mark the fence where the leading edge of the blade will clear the plywood. Then lower the blade and set the fence at the proper distance from the blade. Position the plywood so that the layout line for the lid's crosscut aligns with the mark on the fence. Turn on the saw, and raise the blade until it hits the layout line. Push the plywood into the saw until the blade reaches the second crosscut line. Keeping one hand on the plywood to steady it, carefully reach down and shut off the saw.

Turn the plywood over (top side facing down), and make the crosscuts with a circular saw. Score the layout lines with a knife to minimize chipping. Overcut slightly at the end (it won't show), or finish the cut with a handsaw.

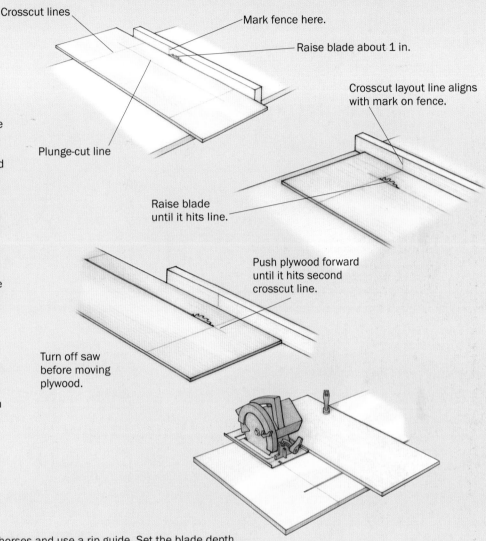

Crosscut lines

Mark fence here.

Raise blade about 1 in.

Plunge-cut line

Crosscut layout line aligns with mark on fence.

Raise blade until it hits line.

Push plywood forward until it hits second crosscut line.

Turn off saw before moving plywood.

Circular-saw Method

Secure the plywood top to sawhorses and use a rip guide. Set the blade depth slightly deeper than the plywood. While holding the guard up, slowly lower the running saw into the plywood by pivoting on the front of the base. Kickback is a possibility, so always stand to the side of the blade and never pull back the saw. Lower the saw and move forward to the far end of the plunge cut.

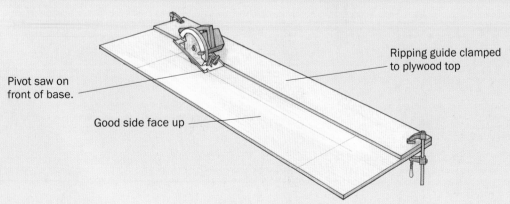

Ripping guide clamped to plywood top

Pivot saw on front of base.

Good side face up

Molding Becomes a Handle for the Lid

In most cases, biscuits aren't necessary to attach a bullnose molding. But because this bullnose is acting as a handle for lifting the window seat's lid, it needs some reinforcement. Using biscuits here makes a clean, strong joint without visible fasteners.

Space biscuits closely for strength. Register the biscuit joiner's fence on the top of the bullnose molding, and cut a series of slots every 6 in. or so.

Protect the molding from the clamps. A scrap of plywood protects the bullnose from the clamps and evenly distributes the clamping pressure. The author marks the time on the lid so that he'll know when the glue has cured. He pulls the clamps after an hour has passed.

Simple Face Frame and Panels Dress Up the Box

The author creates a traditional frame-and-panel look by assembling a face frame with pocket screws and attaching plywood to the back side of it. Molding run around the inside of the panels adds an elegant touch.

A bead of glue ensures a long-lasting, tight joint. Carpenters' glue is spread on the face frame to attach the plywood panel. The brads act like clamps. The plywood panel is held back from the side and bottom edges of the face frame to create a rabbet that engages the window-seat box.

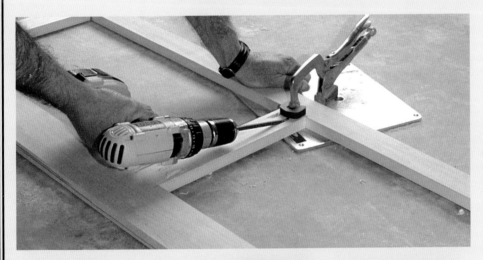

Pocket screws join the face frames from behind. The low angle of the screw draws the joint together without compromising alignment and works faster than biscuits.

Molding dresses up the face frame. After the plywood is attached to the back of the face frame, molding run around the inside creates a traditional frame-and-panel look.

Attach the face frame with nails. The face frame is affixed to the plywood box with 15-ga. finishing nails (or 8ds if you're hand-nailing) and a bead of glue. The top then is nailed around the perimeter, strengthening the seat.

A Piano Hinge Keeps the Lid from Cupping

The lid is attached with a piano hinge to prevent cupping. Its thickness matches the kerf made cutting the lid, so the face of the seat aligns perfectly.

The piano hinge is attached to the seat's lid. The author secures one end of the hinge, then aligns and secures the other end before driving the screws in the middle. Working from one end to the other could mean a crooked installation.

Attaching the lid completes construction. After using the piano hinge as a guide for drilling pilot holes into the hinged edge of the top, the author quickly connects the lid to the window seat.

No Nook? No Problem.

In the project featured in this article, the window seat was built into an existing nook, or alcove. If you don't have a nook, you can create one in a variety of ways. Shown below are two examples, one for a bedroom and one for a den or office.

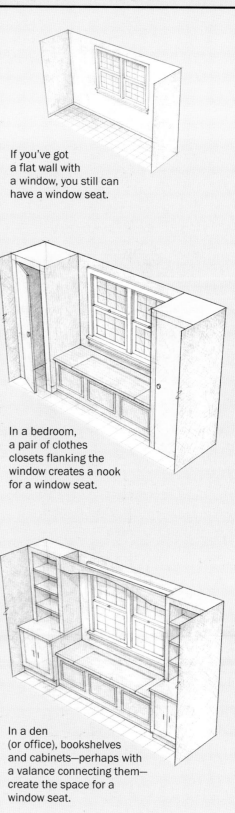

If you've got a flat wall with a window, you still can have a window seat.

In a bedroom, a pair of clothes closets flanking the window creates a nook for a window seat.

In a den (or office), bookshelves and cabinets—perhaps with a valance connecting them—create the space for a window seat.

CREDITS

p. iii: Photo by Roe A. Osborn, courtesy *Fine Homebuilding*, © The Taunton Press, Inc.; p. iv: (left) Photo by Chris Green, courtesy *Fine Homebuilding*, © The Taunton Press, Inc.; (right) Photo by Tom O'Brien, courtesy *Fine Homebuilding*, © The Taunton Press, Inc.; p. v: (left) Photo by Brian Pontolilo, courtesy *Fine Homebuilding*, © The Taunton Press, Inc.; (right) Photo by Tom O'Brien and David Ericson, courtesy *Fine Homebuilding*, © The Taunton Press, Inc.; p. vi: (left) Photo by Gary M. Katz, courtesy *Fine Homebuilding*, © The Taunton Press, Inc.; (right) Photo by Roe A. Osborn, courtesy *Fine Homebuilding*, © The Taunton Press, Inc.; p. 1: (left) Photo by Charles Bickford, courtesy *Fine Homebuilding*, © The Taunton Press, Inc.; (right) Photo by Roe A. Osborn, courtesy *Fine Homebuilding*, © The Taunton Press, Inc.

The articles in this book appeared in the following issues of *Fine Homebuilding*:

p. 4: 10 Rules for Framing by Larry Haun, issue 158. Drawings by Christopher Clapp, courtesy *Fine Homebuilding*, © The Taunton Press, Inc.

p. 13: Avoiding Common Framing Errors by Rick Tyrell, issue 106. Drawings by Christopher Clapp, courtesy *Fine Homebuilding*, © The Taunton Press, Inc.

p. 20: The Well-Framed Floor by Jim Anderson, issue 160. Photos by Chris Green, courtesy *Fine Homebuilding*, © The Taunton Press, Inc.

p. 29: Careful Layout for Perfect Walls by John Spier, issue 156. Photos by Roe A. Osborn, courtesy *Fine Homebuilding*, © The Taunton Press, Inc.; Drawings by Vince Babak, courtesy *Fine Homebuilding*, © The Taunton Press, Inc.

p. 38: All About Headers by Clayton DeKorne, issue 162. Photos by Roe A. Osborn, courtesy *Fine Homebuilding*, © The Taunton Press, Inc.; Drawings by Dan Thornton, courtesy *Fine Homebuilding*, © The Taunton Press, Inc.

p. 46: Mudsills: Where the Framing Meets the Foundation by Jim Anderson, issue 157. Photos by Ron Ruccio, courtesy *Fine Homebuilding*, © The Taunton Press, Inc.; Drawings by Dan Thornton, courtesy *Fine Homebuilding*, © The Taunton Press, Inc.

p. 54: Framing Walls by Scott McBride, issue 82. Photos by Charles Miller, courtesy *Fine Homebuilding*, © The Taunton Press, Inc.; Drawings by Christopher Clapp, courtesy *Fine Homebuilding*, © The Taunton Press, Inc.

p. 65: Framing Gable Ends by Jim Thompson, issue 88. Drawings by Christopher Clapp, courtesy *Fine Homebuilding*, © The Taunton Press, Inc.

p. 72: Straightening Framed Walls by Derek McDonald, issue 133. Photos by Tom O'Brien, courtesy *Fine Homebuilding*, © The Taunton Press, Inc.

p. 78: Framing a Roof Valley by Rick Arnold, issue 161. Photos by Brian Pontolilo, courtesy *Fine Homebuilding*, © The Taunton Press, Inc.; Drawings on pp. 80-81 by Chuck Lockhart, courtesy *Fine Homebuilding*, © The Taunton Press, Inc., Drawings on pp. 82-83 and 85 by Toby Welles/Design Core.

p. 88: A Different Approach to Rafter Layout by John Carroll, issue 115. Photos by Steve Culpepper, courtesy *Fine Homebuilding*, © The Taunton Press, Inc.; Drawings by Dan Thornton, courtesy *Fine Homebuilding*, © The Taunton Press, Inc.

p. 99: Aligning Eaves on Irregularly Pitched Roofs by Scott McBride, issue 91. Photos by Scott McBride, courtesy *Fine Homebuilding*, © The Taunton Press, Inc.; Drawings by Bob Goodfellow, courtesy *Fine Homebuilding*, © The Taunton Press, Inc.

p. 106: Vinyl Siding Done Right by Mike Guertin, issue 149. Photos by Tom O'Brien and David Ericson, courtesy *Fine Homebuilding*, © The Taunton Press, Inc.

p. 115: Installing Horizontal Wood Siding by Felix Marti, issue 96. Photos by Charles Miller, courtesy *Fine Homebuilding*, © The Taunton Press, Inc.; Drawings by Dan Thornton, courtesy *Fine Homebuilding*, © The Taunton Press, Inc.

p. 127: Installing Wood Clapboards by Rick Arnold and Mike Guertin, issue 112. Photos by Andy Engel, courtesy *Fine Homebuilding*, © The Taunton Press, Inc.; Drawings by Rick Daskum, courtesy *Fine Homebuilding*, © The Taunton Press, Inc.

p. 138: 10 Rules for Finish Carpentry by Will Beemer, issue 113. Drawings by Dan Thornton, courtesy *Fine Homebuilding*, © The Taunton Press, Inc.

p. 146: A Simple Approach to Raised-Panel Wainscot by Gary Striegler, issue 165. Photos by James Kidd, courtesy *Fine Homebuilding*, © The Taunton Press, Inc., except p. 153 Photo by Brian Pontolilo, courtesy *Fine Homebuilding*, © The Taunton Press, Inc.; Drawings by Bob LaPointe, courtesy *Fine Homebuilding*, © The Taunton Press, Inc.

p. 154: Crown Molding Around a Cathedral Ceiling by Gary M. Katz, issue 162. Photos by Roe A.Osborn, courtesy *Fine Homebuilding*, © The Taunton Press, Inc.; Drawings by Chuck Lockhart, courtesy *Fine Homebuilding*, © The Taunton Press, Inc.

p. 157: Crown Molding Fundamentals by Clayton DeKorne, issue 152. Photos by Andrew Kline, courtesy *Fine Homebuilding*, © The Taunton Press, Inc., except p. 159 Photo by James Kidd, courtesy *Fine Homebuilding*, © The Taunton Press, Inc. and p. 165 Photo by Tom O'Brien, courtesy *Fine Homebuilding*, © The Taunton Press, Inc.; Drawings by Dan Thornton, courtesy *Fine Homebuilding*, © The Taunton Press, Inc.

p. 167: Building a Fireplace Mantel by Gary M. Katz, issue 93. Photos by Gary M. Katz, courtesy *Fine Homebuilding*, © The Taunton Press, Inc.; Drawings by Christopher Clapp, courtesy *Fine Homebuilding*, © The Taunton Press, Inc.

p. 173: Installing Vinyl-Clad Windows by Rick Arnold and Mike Guertin, issue 129. Photos by Roe A. Osborn, courtesy *Fine Homebuilding*, © The Taunton Press, Inc.

p. 181: New Window in an Old Wall by Rick Arnold, issue 154. Photos by Roe A. Osborn, courtesy *Fine Homebuilding*, © The Taunton Press, Inc.

p. 190: Hanging French Doors by Gary Striegler, issue 160. Photos by James Kidd, Photos by Roe A. Osborn, courtesy *Fine Homebuilding*, © The Taunton Press, Inc., except p. 190 Photo by Rick Green, Photos by Roe A. Osborn, courtesy *Fine Homebuilding*, © The Taunton Press, Inc.; Drawings by Chuck Lockhart, courtesy *Fine Homebuilding*, © The Taunton Press, Inc.

p. 194: A New Door Fits and Old Jamb by Gary M. Katz, issue 165. Photos by Roe A. Osborn, courtesy *Fine Homebuilding*, © The Taunton Press, Inc.; Drawings by Dan Thornton, courtesy *Fine Homebuilding*, © The Taunton Press, Inc.

p. 202: Installing Bifold Doors by Jim Britton, issue 110. Photos by Charles Miller, courtesy *Fine Homebuilding*, © The Taunton Press, Inc.

p. 208: Installing Prehung Doors by Jim Britton, issue 96. Photos by Charles Miller, courtesy *Fine Homebuilding*, © The Taunton Press, Inc.

p. 216: Installing Sliding Doors by Gary M. Katz, issue 108. Photos by Gary M. Katz, courtesy *Fine Homebuilding*, © The Taunton Press, Inc., except pp. 223–225 Photos by Roe A. Osborn, courtesy *Fine Homebuilding*, © The Taunton Press, Inc.

p. 226: Cutting Out Basic Stairs by Eric Pfaff, issue 100. Photos by Roe A. Osborn, courtesy *Fine Homebuilding*, © The Taunton Press, Inc.; Drawings by Christopher Clapp, courtesy *Fine Homebuilding*, © The Taunton Press, Inc.

p. 233: Disappearing Attic Stairs by William T. Cox, issue 89. Photos by Jefferson Kolle, courtesy *Fine Homebuilding*, © The Taunton Press, Inc. except p. 236 Photo by William T. Cox, courtesy *Fine Homebuilding*, © The Taunton Press, Inc.

p. 240: Installing Handrails Made from Stock Parts by Lon Schleining, issue 121. Photos by Charles Bickford, courtesy *Fine Homebuilding*, © The Taunton Press, Inc., except p. 241 Photo by Scott Phillips, courtesy *Fine Homebuilding*, © The Taunton Press, Inc.

p. 249: A Quick Way to Build a Squeak-Free Stair by Alan Ferguson, issue 104. Photos by Sharon Mills, courtesy *Fine Homebuilding*, © The Taunton Press, Inc.; Drawings by Kathleen Rushton, courtesy *Fine Homebuilding*, © The Taunton Press, Inc.

p. 260: A Built-In Hardwood Hutch by Stephen Winchester, issue 85. Photos on p. 261 (bottom) and p. 267 by Stephen Winchester, courtesy *Fine Homebuilding*, © The Taunton Press, Inc.; Photos on p. 261 (top) and p. 266 by Rich Ziegner, courtesy *Fine Homebuilding*, © The Taunton Press, Inc.

p. 268: Designing and Building an Entertainment Center by Brian Wormington, issue 120. Photos by Andy Engel, courtesy *Fine Homebuilding*, © The Taunton Press, Inc., except p. 272 (inset) Photo by Scott Phillips, courtesy *Fine Homebuilding*, © The Taunton Press, Inc.; Drawings by Vince Babak, courtesy *Fine Homebuilding*, © The Taunton Press, Inc.

p. 274: Bookshelf Basics by Bruce Greenlaw, issue 103. Photos by Bruce Greenlaw, courtesy *Fine Homebuilding*, © The Taunton Press, Inc., except p. 275 courtesy Thos. Moser Cabinetmakers.

p. 283: Installing Kitchen Cabinets by Kevin Luddy, issue 132. Photos by Roe A. Osborn, courtesy *Fine Homebuilding*, © The Taunton Press, Inc.

p. 292: Simple Frameless Cabinets Built on Site by Joseph B. Lanza, issue 138. Photos by Scott Gibson, courtesy *Fine Homebuilding*, © The Taunton Press, Inc.; Drawings by Dan Thornton, courtesy *Fine Homebuilding*, © The Taunton Press, Inc.

p. 298: Enhance Any Room with a Window Seat by Gary Striegler, issue 158. Photos by James Kidd, courtesy *Fine Homebuilding*, © The Taunton Press, Inc., except Photo on p. 299 by Rick Green, courtesy *Fine Homebuilding*, © The Taunton Press, Inc. and Photo on p. 307 by Scott Phillips, courtesy *Fine Homebuilding*, © The Taunton Press, Inc.; Drawings by Vince Babak, courtesy *Fine Homebuilding*, © The Taunton Press, Inc.

INDEX

Taunton's FOR PROS BY PROS Series
A Collection of the best articles from Fine Homebuilding magazine

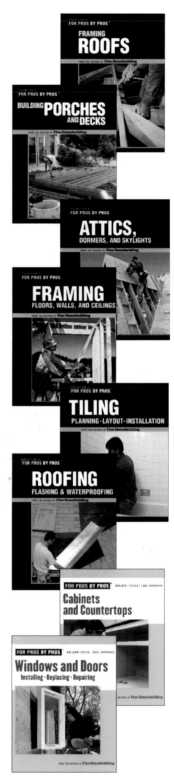